Operational Amplifier
Circuit Manual

Operational Amplifier Circuit Manual

Robert J. Traister

ACADEMIC PRESS, INC.
Harcourt Brace Jovanovich, Publishers
San Diego New York Berkeley Boston London Sydney Tokyo Toronto

Academic Press, Inc.
San Diego, California 92101

United Kingdom Edition published by
Academic Press Limited
24–28 Oval Road, London NW1 7DX

Library of Congress Cataloging-in-Publication Data

Traister, Robert J.
    Operational amplifier circuit manual / Robert J. Traister
      p.   cm.
    Includes index.
    ISBN   0-12-697405-5   (alk. paper)
    1. Operational amplifiers--Handbooks, manuals, etc. 2. Integrated circuits--Handbooks, manuals, etc.   I. Title
    TK7871.58.O6T73   1989
    621.3815'35--dc19   88-35594
                                                               CIP

Printed in the United States of America
89 90 91 92   9 8 7 6 5 4 3 2 1

This book is dedicated to my uncles, Charles Traister and Jim Colbert, both of whom have had a profound impact on my life, though they probably don't realize it and may not have seen any evidence of it.

Contents

Chapter 9
Op Amp Filter Circuits   119

Chapter 10
Digital Applications   123

Chapter 11
Miscellaneous Circuits   129

Circuit Index   155

Preface

Operational amplifier circuits: These three words describe the great majority of the contents of this text. Circuits for engineers, circuits for technicians, and circuits for many and varied applications are included.

This book is a convenient, generic reference source for electronic circuits that utilize IC operational amplifiers as their major or core components. The proliferation of highly efficient op amps is only outnumbered by uncountable circuits that incorporate them. This, then, is a collection of circuits deemed to be representative of the many useful circuits put forth by the major electronic manufacturers. A few circuits are included for which little use has been found, but they are presented along with the rest for those readers with special circuit requirements.

Each circuit is accompanied by a terse explanation of its operation and/or suggestions for possible uses, modifications, and hints that may aid the engineer or technician who is developing a specialized project.

While circuits were not chosen with any particular criterion in mind regarding ease of construction, availability, cost, and so on, most of those found here are eminently practical and do not require any exotic, hard-to-find components or unusual construction practices. Nor will many of these circuits be constructed on a literal basis by most readers. Rather, they are offered as building blocks for more elaborate, complex circuits, many of which exist only in the mind of the user.

The reader in need of a particular type of circuit that incorporates the modern op amp as the basis for its operation will probably find it, or one very close to it, in these pages. Op amp circuit designs reach into every field of electronics, from the most stringent government, commercial, or military uses to the amateur and hobby spectrum. It is hoped that this collection of op amp circuits will serve as both a ready reference source depicting what the field currently has to offer and a practical schematic for actual circuit design and assembly.

Robert J. Traister

Chapter 1

The Operational Amplifier

The most commonly utilized integrated circuit (IC) is the operational amplifier or op amp. The op amp has been in existence for many decades, starting as a highly complex device comprised of vacuum tubes and a myriad of other discrete components and utilized in the now-primitive analog computers of a bygone era. Even with the coming of the solid-state genre of electronics, the operational amplifier remained a collection of discrete devices. While the transistor-based version replaced its vacuum tube counterpart, the op amp did not really come into prominence until after it was integrated into a single, small chip.

Certainly, the cost of a typical operational amplifier constructed from discrete components was a large factor in its lack of popularity in early electronic designs. However, with improved fabrication techniques coupled to large volume production, the modern monolithic operational amplifier is truly a building block for many different types of complex circuits that address, literally, every imaginable field of linear applications.

Put simply, the operational amplifier is a DC-coupled differential amplifier offering very high gain. The ideal op amp (theoretical) would possess infinite voltage gain, zero output impedance, infinite input impedance, infinite bandwidth, zero noise, and zero temperature drift. Naturally, this theoretical device is unapproachable in a physical contrivance, but close proximities of infinite voltage gain, zero output impedance, and infinite input impedance are achieved regarding practical definitions of these infinite quantities. However, when it comes to bandwidth, noise, and drift, practical devices of today fall quite a bit short of the theoretical "ideal" amplifier.

With input impedance at a very high value, an amplifier responds to the signal voltage without drawing appreciable current from the source. A very low output impedance allows for a constant voltage value across the output, regardless of the current being drawn by the load.

The general utility associated with the operational amplifier hinges on the fact that it is intended for use in a feedback loop whose properties determine the feed-forward characteristics of the amplifier–loop combination.

In electronics, we usually think of access points to a circuit as terminals, and these terminals are usually referenced to ground in some way or another. Polarity designations form the brunt of such ground references. In an operational amplifer, the input (and often the output) is not committed in regard to polarity. Rather, these circuit entry or exit points are established by a particular circuit use or connection.

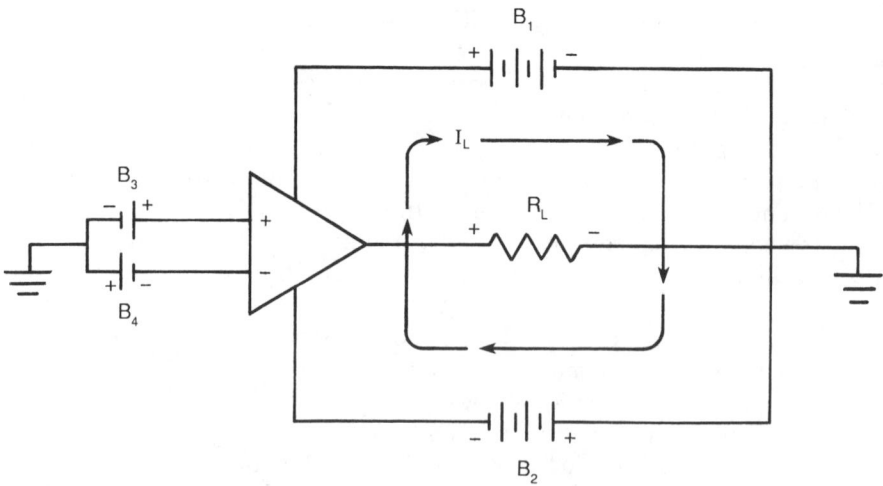

Fig. 1. Input signal causes the amplifier to sink the load current.

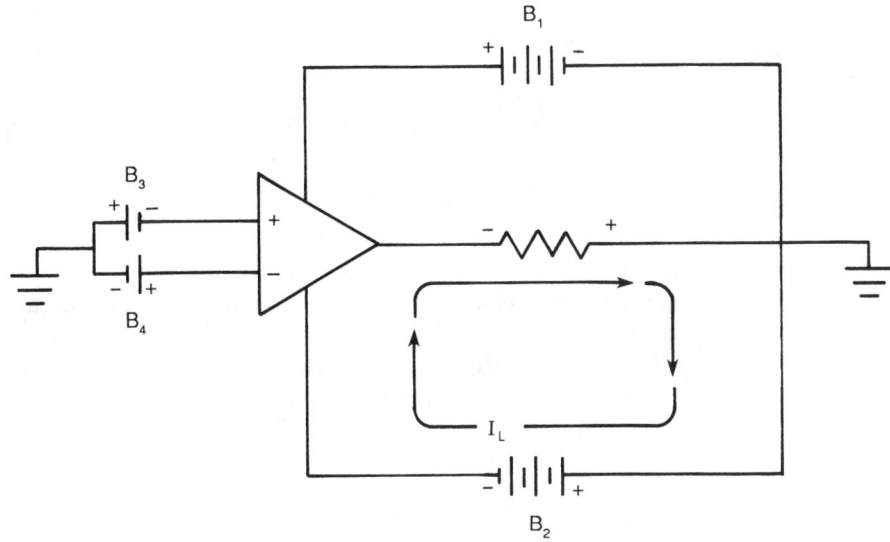

Fig. 2. Input signal causes the amplifier to source the load current.

Figure 1 shows a simple op amp configuration that uses two bipolar power sources. B_1 and B_2 bias the amplifier and supply the load current. The common reference point for these power sources is the op amp's output terminals. The input signal is generated by B_3 and B_4 which are referenced to the common output terminal. This schematic shows that a positive voltage in reference to ground is applied to the noninverting ($+$) input terminal. This means that the high side of the output will go positive and sink the load current. The negative potential applied to the inverting ($-$) input causes the output to be positive.

Figure 2 shows a similar configuration, however the input signal polarity is reversed. The negative-going signal applied at the noninverting input terminal and positive signal at the inverting terminal cause the output to go negative and to source the load current.

Both conditions (sinking and sourcing) occur in these two examples because the input polarities remain bipolar. However, if only one input signal source is reversed, then the inputs would oppose rather than aid each other, creating a condition where the output voltage would be zero. When the input signals are of different potentials, the output is proportional to this difference and thus the term "differential amplifier."

Definition of Terms

In discussing operational amplifiers, it is necessary to clearly define the terms that are used to describe the various parameters of circuit operation. The following mini-glossary should be helpful to the reader in this regard.

Bandwidth: That frequency at which the voltage gain of an operational amplifier is reduced to 0.707106 times the low frequency value. (Note: This multiplier is theoretically stated as the quantity 1 divided by the square root of 2, thus 0.707106. . . .)

Common-mode rejection ratio: The ratio of the input common-mode voltage range to the peak-to-peak change in input offset voltage over this range.

Harmonic distortion: That percentage of distortion defined as 100 times the ratio of the rms sum of the harmonics to the fundamental. Percentage of harmonic distortion is equal to

$$\frac{(V_2^2 + V_3^2 + V_4^2 + V_5^2 + \cdots)^{1/2} \times 100\%}{V_1}$$

where V_1 is the rms amplitude of the fundamental and V_2, V_3, V_4, V_5, and so on are the rms amplitudes of the individual harmonics.

Input bias current (IBC): The average of the two input currents.

Input common-mode voltage range: The range of voltages on the input terminals for which the amplifier is operational. The specifications are not guaranteed over the entire full common-mode voltage range unless specifically stated.

Input impedance: The ratio of input voltage to input current under the stated conditions for source resistance (R_s) and load resistance (R_l).

Input offset current: The difference in the currents into the two input terminals at the instant the output is at zero.

Input offset voltage: The voltage that must be applied between the input terminals through two equal resistances to obtain an output potential of zero.

Input resistance: The ratio of the change in input voltage to the change in input current on either input with the other grounded.

Input voltage range: The range of voltages on the input terminals for which the amplifier operates within stated specifications.

Large-signal voltage gain: The ratio of the output voltage swing to the change in input voltage required to drive the output from zero to this voltage value.

Offset voltage temperature drift: The average drift rate of offset voltage for a thermal variation from room temperature to the indicated temperature maximum.

Output impedance: The ratio of output voltage to output current under stated conditions for source resistance (R_s) and load resistance (R_l).

Output resistance: The small signal resistance seen at the output with the output voltage near zero.

Output voltage swing: The peak output voltage swing, referred to zero, that can be obtained without clipping.

Power supply rejection: The ratio of the change in input offset voltage to the change in the power supply voltages producing it.

Settling time: The time between the initiation of the input step function and the time when the output voltage has settled to within a specified error band of the final output voltage.

Slew rate: The internally limited rate of change in output voltage with a large-amplitude step function applied to the input.

Supply current: The current required from the power supply to operate the amplifier with no load and the output midway between the supplies.

Transient response: The closed-loop step-function response of the amplifier under small-signal conditions.

Voltage gain: The ratio of output voltage to input voltage under the stated conditions for source resistance (R_s) and load resistance (R_l).

Intensive development of the operational amplifier, especially in integrated form, has resulted in circuits which are excellent engineering approximations of the ideal (theoretical) op amp discussed earlier. Quantity prices for the best contemporary integrated amplifiers are quite low when compared with transistor prices of less than a decade ago. The low cost and high quality of op amps allows them to be utilized in equipment whose existence would not be practical if discrete components were the basic building block units.

The extensive use of the op amp in wide-ranging circuitry today has to do with the black box concept which completely disregards the internal makeup of the chip. The internal workings, in general, can be ignored when such a device is operated in a linear manner.

Figure 3 shows the pin configuration of one of the most common types of op amps, the 741. Circuit connection points of main interest to the electronics designer include the inverting and noninverting terminals, output terminal, and the positive and negative power supply terminals. The two offset null terminals are used in a circuit

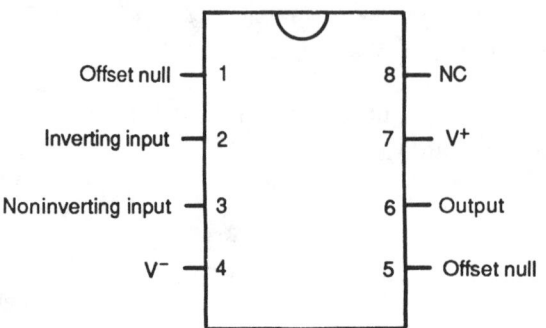

Fig. 3. Pin configuration of a 741 IC operational amplifier.

external to the op amp proper, and are utilized to help compensate for problems that arise due to age or to fabrication techniques. Minimum use of the offset null can be obtained with the proper choice of components that are connected to the inverting and noninverting inputs.

Figure 4 shows the reference parameters used to describe operating properties. All potential values are considered as voltage rises from the common node. A positive power supply is connected to the $V+$ terminal. The negative supply goes to $V-$. V_o is the voltage between the output terminal and the common node; V_a is measured between the inverting terminal and the common node; and V_b is measured between the noninverting terminal and the common node.

The currents associated with each node are also identified in Fig. 4. Using Kirchhoff's law, the sum of all currents entering the operational amplifier will be zero as in

$$I_a + I_b + I_o + I_{plus} + I_{minus} = 0$$

A common mistake often made when viewing modified operational amplifier drawings without the power supply connections shown is to assume that $I_a + I_b = -I_o$. This is generally not true.

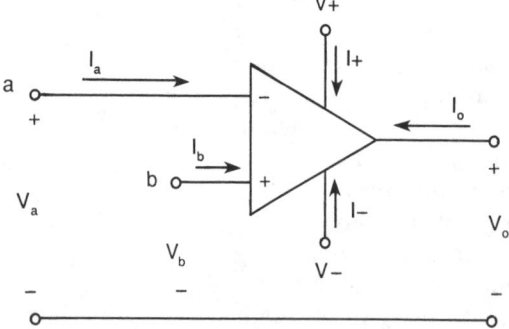

Fig. 4. Previous op amp circuit with node identification.

Figure 5 shows a representation of an ideal operational amplifier. Here, R_i is the input resistance between the inverting and the noninverting terminals. This value is typically very large, ranging from a minimum of approximately 1 MΩ to a maximum of 5 MΩ or more. The voltage gain A is positive and typically is large, being on the order of 15,000 to over 100 million. Finally, the output resistance R_o is very small with typical values of less than 80 Ω.

Again referring to Fig. 5, open circuit output voltage V_o is defined by

$$V_o = A(V_b - V_a)$$

The output voltage is proportional to the voltage differences between the inverting and noninverting terminals. The open-loop gain A is the gain that the op amp will exhibit when a feedback network is not utilized. For the above question to be true, the operational amplifier must be operating within its linear region. This linearity is described by

$$V- \leq V_o \leq V+$$

As an example, the 741 op amp when operated with 15 V($V+$) has a typical open-loop gain of 200,000. This is calculated by

$$15 \geq 200,000 \, (V_b - V_a)$$
$$0.000075 \geq (V_b - V_a)$$

The relations stated here would indicate that under typical operating conditions V_b and V_a must not differ by more than approximately 75 μV in order for the op amp to maintain operation within the linear portion of its curve. Stated practically, this means that

$$V_a = V_b$$

This requirement of a small or nonexistent difference between the two input terminals results in the frequently used concept of the virtual ground. Even though the resistance between the two terminals

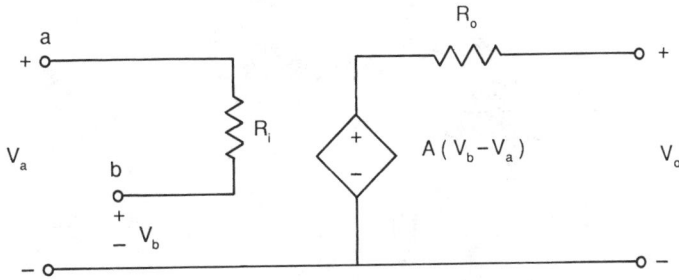

Fig. 5. The ideal operational amplifier.

is often in the megohm range, designers picture a virtual ground between the two. This means that, although there is no actual connection or short circuit between the two, they are at the same voltage level for a potential difference of zero.

Figure 6 shows another equivalent model of an operational amplifier. This one is connected with feedback circuitry to form a noninverting amplifier. This type of amplifier produces an output potential which is a multiple of the input potential and is of the same polarity. Since the open-loop gain A is positive, V_a is less than V_b and V_o is less than $2V_b$. This leads to the following approximation.

$$V_a = V_b$$
$$V_o = 2V_b$$

This means that for all practical purposes V_a does equal V_b, and this allows the input resistance R_i to be replaced by an open. Since the voltage at the inverting terminal V_a is now known, the voltage (V_o) which results in this value can be calculated. To further simplify this process, it can be assumed that the voltage across the output V_o emanates directly from the dependent voltage source and that the output resistance R_o is effectively zero. These two assumptions lead to the approximate equivalent model of the op amp connected in a noninverting configuration as shown in Fig. 7.

This model would seem to present a contradiction in that if $V_a = V_b$, then $V_b - V_a$ must be equal to zero. Therefore, the output would always be zero. However, it must be understood that V_o is equal to $2V_b$ if V_a is equal to V_b. It is most practical to simply remember that this is an approximate solution and by setting V_a equal to V_b, the

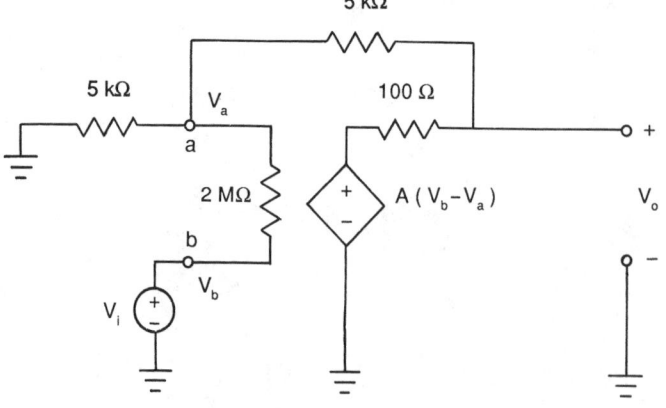

Fig. 6. Op amp equivalent model.

Fig. 7. Noninverting amplifier equivalent.

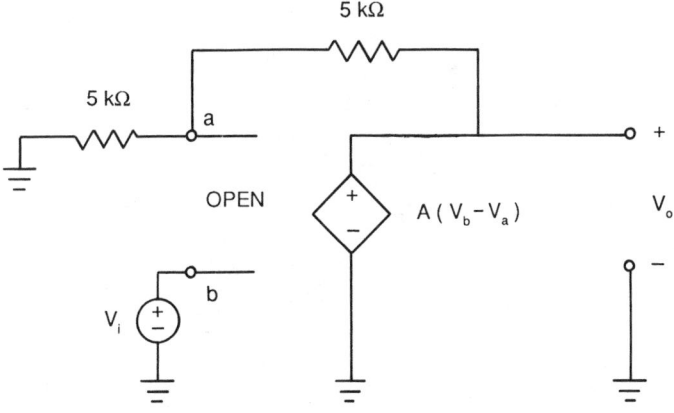

output voltage V_o can quickly be determined. The differences between the true solution and the approximate solution using this last model become much smaller as the open-loop gain A becomes larger. As the open-loop gain approaches infinity, the differences approach equality.

It was mentioned earlier that the proper choice of resistors connected at the input terminals would minimize the need for adjustments to the offset null. Such adjustments are needed when voltages at the input terminals are not within very close tolerance of being equal during the time when the input signal voltage is zero. While under full operating conditions, it is assumed theoretically that there is zero current flow through the input terminals. This is the theoretical assumption, however it is incorrect from a true operational point of view. For the op amp to work an input bias current must exist at the input terminals.

As an example, the input bias for the 741 op amp is typically 30 nA. While there is a very small current value, it must be taken into account when designing the input circuitry in order for voltage at the input terminals to be nearly equal. To accomplish this it is necessary to make admittances at the input terminals equal, which causes the output voltage V_o to fall to zero under conditions of no input signal.

Figure 8 shows an input circuit design that minimizes the need for an offset null adjustment. A 2500-Ω resistor is placed in series with the noninverting input terminal so that the admittance to ground from this terminal is equal to the admittance to ground at the inverting terminal. A further discussion of this principle is found in the next section of this chapter.

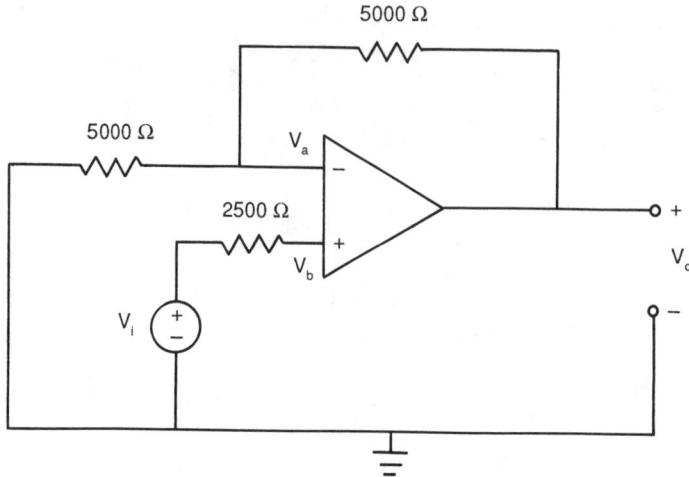

Fig. 8. Using a series resistor to minimize offset null adjustment.

The Inverting Amplifier

The basic operational amplifier assembled in an inverting amplifier configuration is shown in Fig 9. This circuit exhibits a closed-loop gain equivalent to R2/R1 providing that this ratio is small in comparison to the amplified open-loop gain. The input impedance is equal to R1, and the closed-loop bandwidth is equal to the unity gain frequency divided by the quantity of one plus the closed-loop gain. This amplifier produces an output voltage which is a negative multiple of its input voltage value.

Referring to Fig. 9 again, R3 should be chosen to be equal to the parallel equivalent of R1 and R2. This is necessary in order to minimize the offset voltage error due to bias current and so there will be an offset voltage at the amplifier output which is equal to the closed-

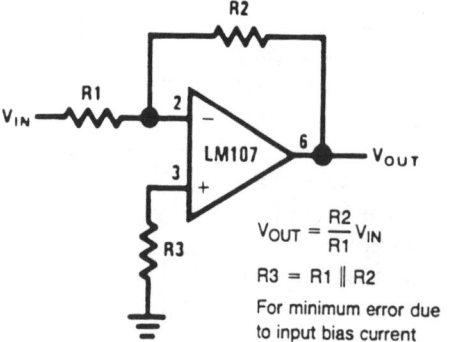

Fig. 9. Inverting amplifier. —Courtesy of National Semiconductor Corporation, Santa Clara, California.

loop gain times the offset voltage at the amplifier input. The input offset voltage is fixed for any particular amplifier (i.e., device), but the contribution due to input bias current is dependent on the circuit configuration used. For minimum offset voltage at the amplified input without circuit adjustment the source resistance for both inputs must be equal, as was previously discussed. In this case the maximum offset voltage would be the algebraic sum of the amplifier offset voltage and the voltage drop across the source resistance due to offset current. Amplifier offset voltage is the predominant error term for low source resistances, and offset current causes the main error for high source resistances.

In applications where high source resistance is the configuration, the offset voltage at this amplifier may be adjusted by altering the value of R3 and using the variation in voltage drop across this component as an input offset voltage trim. Offset voltage at the amplifier output is not as important in AC-coupled applications, where the prime consideration is that any offset voltage (at the output) reduces the peak-to-peak linear output swing of the amplifier.

The gain/frequency characteristic of the inverting amplifier and its feedback network must be designed to prevent oscillation. To accomplish this goal it is necessary that the phase shift through the amplifier and the feedback network never exceed 180°. This applies to any frequency where the gain of the amplifier and its feedback network is greater than one. From a practical standpoint, the phase shift should not even approach 180°, because this in itself is a condition of considerable instability. When the attenuation of the feedback network is at or near zero, the most critical circuit conditions occur.

Figure 10 is used to describe the various parameters and their relationships within a common inverting amplifier configuration. This discussion assumes that this schematic represents the perfect or theoretically ideal amplifier. Summing the currents exiting the V_a node produces the following equations.

$$\text{(A)} \quad (0 - V_i)/R + (0 - V_o)/R_f = 0$$
$$\text{(B)} \quad -V_i/R = V_o/R_f$$
$$\text{(C)} \quad (-R_f/R)V_i = V_o$$
$$\text{(D)} \quad -R_f/R = V_o/V_i$$

Equation (C) states the relationship between the input voltage V_i and the output voltage V_o. This amplifier produces an output voltage which is a negative multiple of the input voltage value. The output voltage values range from smaller than to larger than the input voltage. The actual value will depend upon the ratio of R_f to R. This operation is totally different from that of the noninverting amplifier which has an output voltage whose magnitude is always larger than the input voltage. Equation (D) is the transfer function for the inverting amplifier.

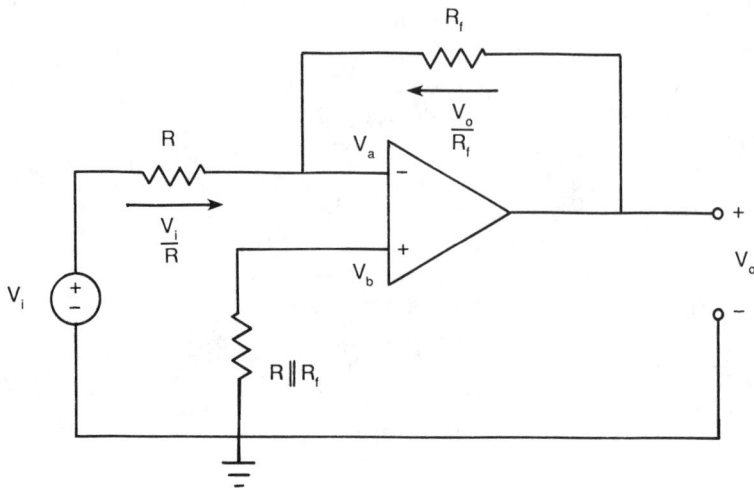

Fig. 10. Inverting amplifier parameter designations.

The Noninverting Amplifier

The noninverting amplifier configuration shown in Fig. 11 exhibits a high input impedance and gives a closed-loop gain that is equal to the sum of R1 and R2 divided by R1. The closed-loop 3-dB bandwidth is equal to the amplifier unity gain frequency divided by the closed-loop gain.

Naturally, a main difference of this circuit as compared to the previous one is that it produces an output that is not inverted. Also, the input impedance is very high and is equal to the differential input impedance multiplied by the loop gain. In DC-coupled applications input impedance is not as important as input current and the voltage drop across the source resistance that it creates.

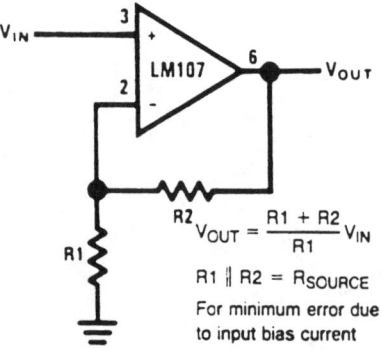

Fig. 11. Noninverting amplifier. —Courtesy of National Semiconductor Corporation, Santa Clara, California.

Unlike the inverting amplifier, the output of this noninverting configuration will go into saturation should the input be allowed to float. This is an important consideration if the noninverting amplifier is to be switched from source to source.

Figure 12 shows a schematic drawing similar to the previous one. This drawing is representative of the ideal (theoretical) inverting amplifier whereby $V_a = V_b$, and the open-loop gain A and the input resistance are infinite.

It can be seen that the voltage at the noninverting (+) terminal V_b is equal to the input voltage V_i, owing to the op amp's infinite input resistance R_i. The input voltage value of V_i appears at the inverting terminal due to the virtual ground which exists between the two input terminals as discussed earlier. Summing the currents at the inverting terminal produces the following equations.

(A) $V_a = V_i$
(B) $V_i/R + (V_i - V_o)/R_f = 0$
(C) $V_i/R = (V_o - V_i)/R_f$
(D) $(R_f/R)V_i = V_o - V_i$
(E) $(R_f/R)V_i + V_i = V_o$
(F) $((R_f + R)/R)V_i = V_o$
(G) $(R_f + R)/R = V_o/V_i$

Equation (A) describes the relationship between the output voltage V_o and the input voltage V_i, while Eq. (G) is the transfer function for the circuit. This discussion assumes that the op amp is indeed operating within the linear portion of its curve such that the magnitude of the output voltage does not exceed that of the power supply voltage.

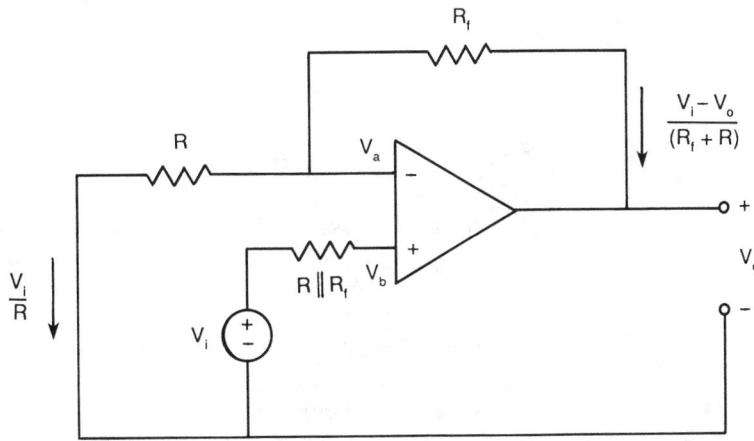

Fig. 12.   The ideal inverting amplifier.

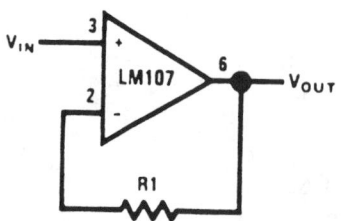

Fig. 13. Unity gain buffer. —Courtesy of National Semiconductor Corporation, Santa Clara, California.

Unity Gain Buffer

The op amp circuit that exhibits the highest input impedance is the unity gain buffer shown in Fig. 13. The actual input impedance value is a product of the differential input impedance and the open-loop gain in parallel with common mode input impedance. The gain error of the unity gain buffer circuit is equal to the reciprocal of the amplifier open-loop gain or to the common-mode rejection, whichever is less.

Actually, input impedance is a misleading concept in DC-coupled unity gain buffers. The source resistance supplies the bias current and will cause an error at the amplifier input, owing to its voltage drop across this resistance. Due to this trait, a low bias current op amp makes the best candidate for a unity gain buffer when working from high source resistances.

In designing a unity gain buffer it is mandatory that the amplifier be compensated for unity gain operation. The output swing of the amplifier may be limited by the common mode range, as some op amps exhibit a latch-up mode when the amplifier common mode is exceeded.

Summing Amplifier

A summing amplifier is a special inverting amplifier that produces an inverted output equal to the weighted algebraic sum of all three inputs. As shown in Fig 14, the gain of any input is equal to the ratio of the appropriate input resistor to the feedback resistor (R4). Bandwidth is calculated in the same manner as for the inverting amplifier discussed earlier. This circuit offers the advantage of no interaction between inputs, and operations such as summing and weighted averaging are implemented in an easy manner.

Fig. 14. Summing amplifier.—
Courtesy of National Semiconductor
Corporation, Santa Clara, California.

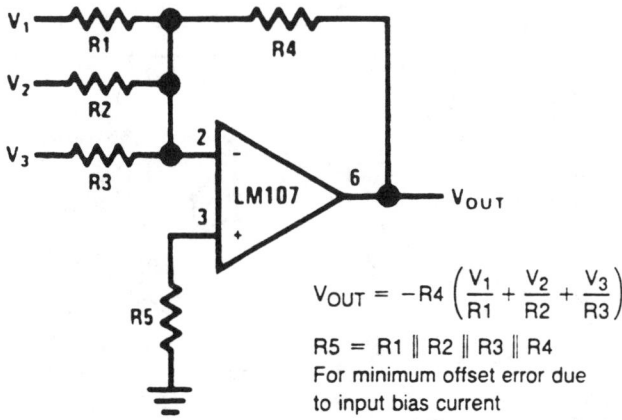

$$V_{OUT} = -R4 \left(\frac{V_1}{R1} + \frac{V_2}{R2} + \frac{V_3}{R3} \right)$$

R5 = R1 ∥ R2 ∥ R3 ∥ R4
For minimum offset error due
to input bias current

Difference Amplifier

The complement of the summing amplifier is called a difference amplifier and allows for the subtraction of two voltages or for the cancellation of a signal common to the two inputs. Figure 15 shows an ideal difference amplifier.

As with the summing amplifier, bandwidth may be calculated in the same manner as for a standard inverting amplifier; however, input impedance for the two inputs is not necessarily equal.

Fig. 15. Ideal difference amplifier.—
Courtesy of National Semiconductor
Corporation, Santa Clara, California.

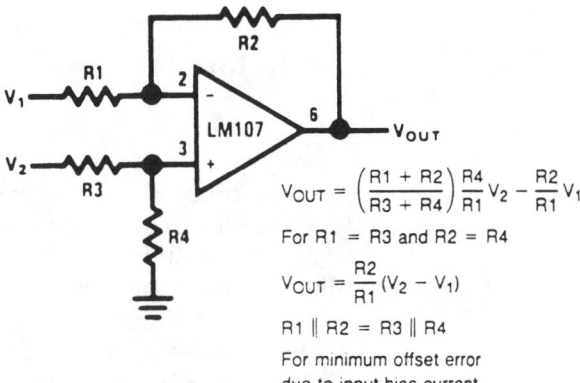

$$V_{OUT} = \left(\frac{R1 + R2}{R3 + R4} \right) \frac{R4}{R1} V_2 - \frac{R2}{R1} V_1$$

For R1 = R3 and R2 = R4

$$V_{OUT} = \frac{R2}{R1} (V_2 - V_1)$$

R1 ∥ R2 = R3 ∥ R4

For minimum offset error
due to input bias current

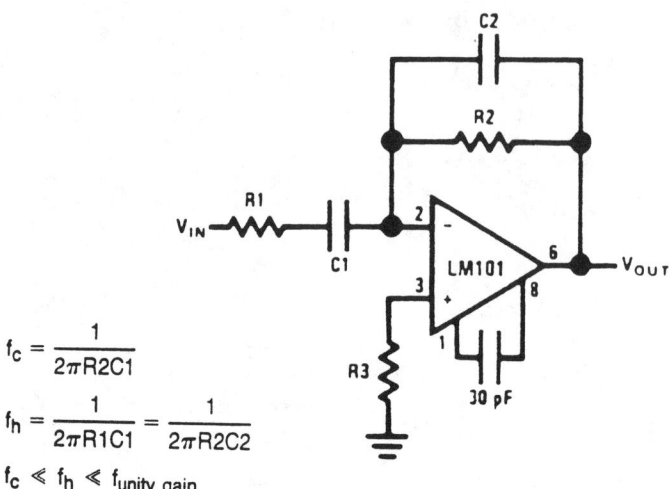

Fig. 16. Practical differentiator.—Courtesy of National Semiconductor Corporation, Santa Clara, California.

$$f_c = \frac{1}{2\pi R2C1}$$

$$f_h = \frac{1}{2\pi R1C1} = \frac{1}{2\pi R2C2}$$

$$f_c \ll f_h \ll f_{unity\ gain}$$

Differentiator

A differentiator as shown in Fig. 16 performs the mathematical operation of differentiation. This is a practical differentiator circuit which does not exactly follow the mathematical form, because this would not be practical due to its susceptibility to high-frequency noise. In the circuit diagram, stability and noise problems are overcome by R1 and C2 which are not included in the mathematical model. C2 and R2 form a 6 dB per octave high-frequency roll-off in the feedback network. R1 and C1 form the same type of network at the input. This creates a total high-frequency roll-off of 12 dB per octave, thus reducing the effect of high-frequency input and amplifier noise.

Integrator

The circuit shown in Fig. 17 performs the mathematical integration operation. This circuit is really a low-pass filter with a frequency response decreasing at 6 dB per octave. It is necessary for this circuit to be provided with an external method of establishing initial conditions. This is illustrated here as S1. When this switch is in position 1, the amplifier is connected in unity gain, and C1 is discharged creating an initial condition of zero volts. When the switch is in the second position, the amplifier is connected as an integrator and its output will change in accordance with a constant multiplied by the integral of the input voltage.

Fig. 17. Integrator.—Courtesy of National Semiconductor Corporation, Santa Clara, California.

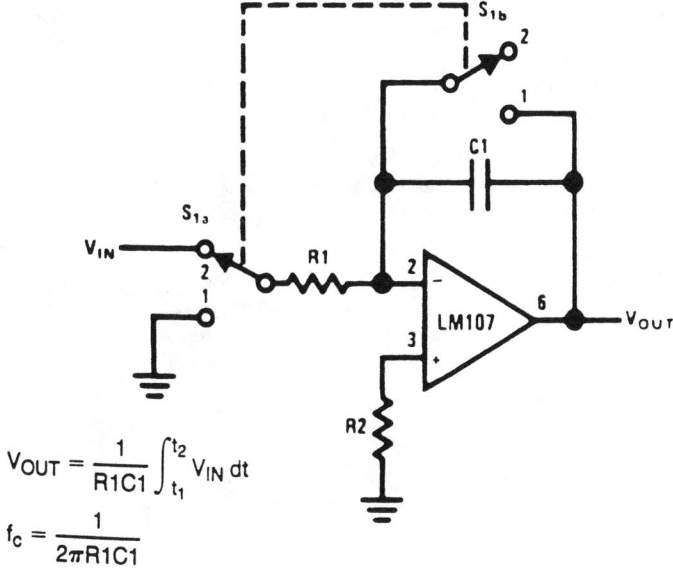

$$V_{OUT} = \frac{1}{R1C1} \int_{t_1}^{t_2} V_{IN} \, dt$$

$$f_c = \frac{1}{2\pi R1C1}$$

Drift Compensation Techniques

The drift performance in DC amplifiers can often be improved by using adjustable circuit components. The simplest and most effective method of compensating for bias currents is shown in Fig. 18. Here, the offset produced by the bias current at the inverting input of the

Fig. 18. Summing amplifier bias current compensation for a fixed source resistance.—Courtesy of National Semiconductor Corporation, Santa Clara, California.

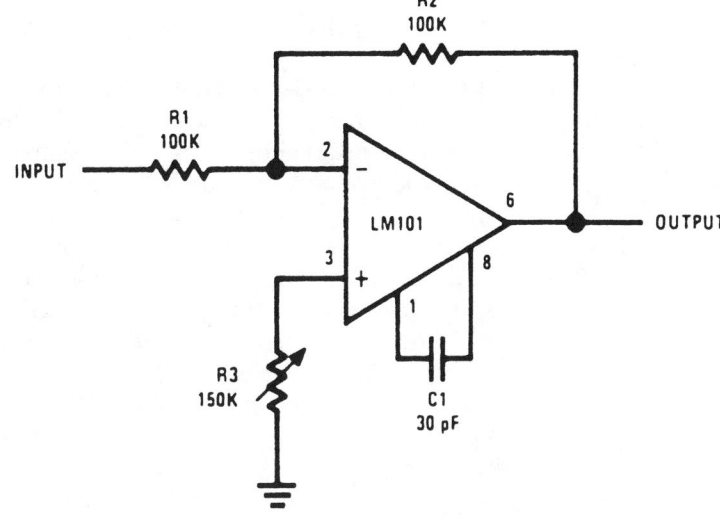

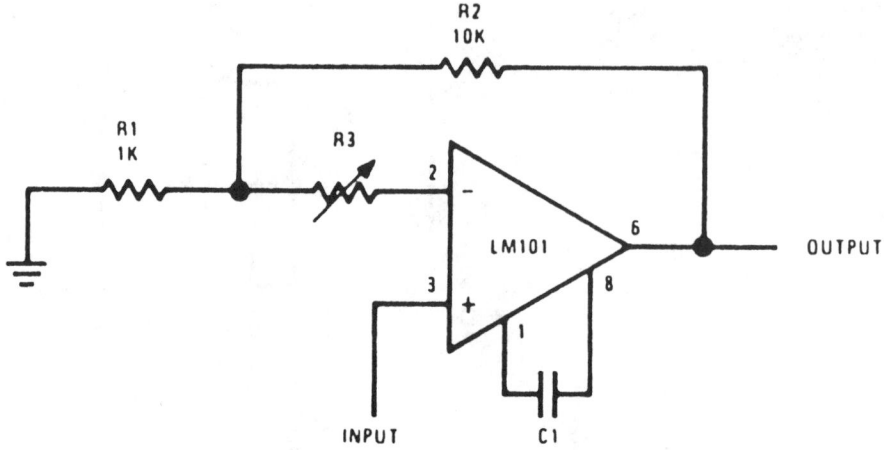

Fig. 19. Noninverting amplifier with bias current compensation.—Courtesy of National Semiconductor Corporation, Santa Clara, California.

op amp is canceled by the offset voltage produced across the variable resistor R3. This is an extremely simple compensation scheme, but there are problems associated with such simplicity. One of these lies in the fact that any given compensation setting works only with fixed feedback resistors. Therefore, the variable control must be readjusted if the equivalent parallel resistance of R1 and R2 changes.

Figure 19 shows a similar compensation arrangement for a noninverting amplifier where the variable resistance is located between the noninverting input and ground. The offset voltage produced across the DC resistance of the source due to the input current is canceled by the drop across R3. In this usage R3 must have a maximum value that is approximately 3 times the ohmic value of the source resistance. The equivalent parallel resistance of R1 and R2 should be less than one-third the input source resistance. Again, this is a simple scheme that exhibits the same deficiencies as did the previous circuit.

To partially overcome the problem of adaptation to varying DC source resistance values, the configuration shown in Fig. 20 is often utilized. Here a current is injected into the input terminal from the base of a bipolar transistor. This is a P-N-P type, since N-P-N transistors are used at the input on the integrated amplifier on the chip. The base current of the P-N-P transistor balances out the base current of the N-P-N.

Fig. 20. Alternate compensation
method for summing amplifier.—
Courtesy of National Semiconductor
Corporation, Santa Clara, California.

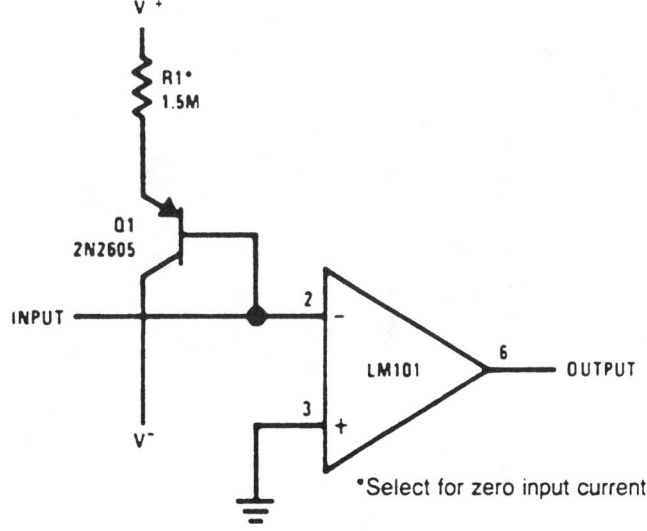

This arrangement works well in summing amplifiers as shown, but
it does have limitations (sometimes severe) in other applications.
One of these limitations is a severe reduction in input impedance
when the base of the transistor is connected to the noninverting input
as would be the case with a voltage follower. To overcome this diffi-
culty the circuit shown in Fig. 21 is often used. Here the emitter of
the P-N-P transistor is fed from a current source so that the compensating current does not vary with the input voltage value.

Another simpler scheme is shown in Fig. 22 to address the same
concern. The compensating current is obtained through a resistor
connected across a diode which is in turn bootstrapped to the output.
In this configuration, the diode acts as a regulator so that the compensating current does not change appreciably with signal level. This
provides input impedances of approximately 1000 MΩ but the temperature compensation is not as good as with the previous compensation circuit.

All of the previous circuits have been designed for specific applications. The one depicted in Fig. 23 is a general scheme that compensates both inputs over the full common-mode range. In addition to
current compensation, this technique also compensates the amplifier
for power supply and temperature variations. This circuit is suitable
for use in summing amplifiers and in noninverting amplifiers.

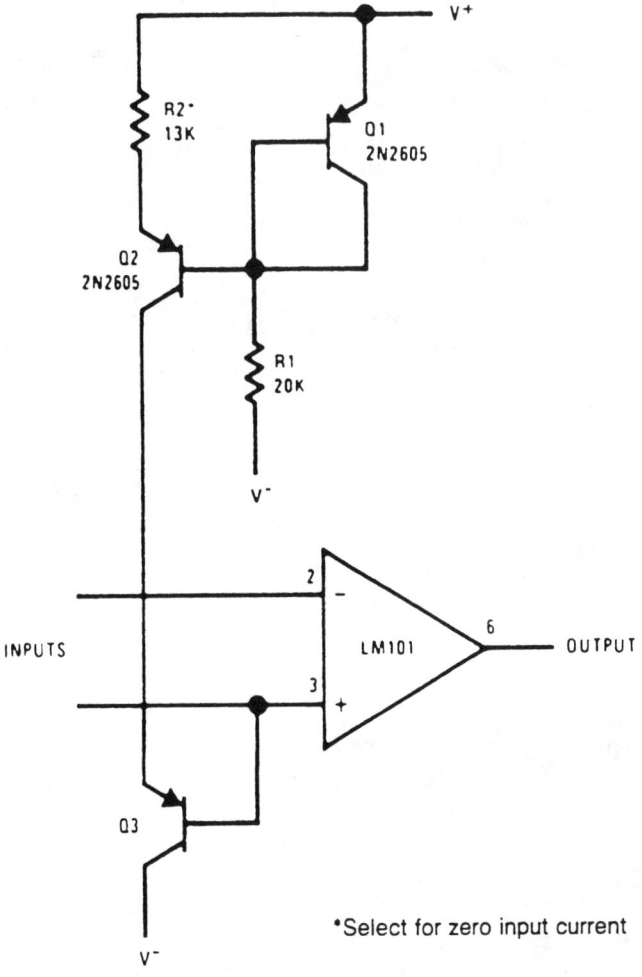

*Select for zero input current

Fig. 21. Noninverting amplifier bias current compensation over large common-mode range.

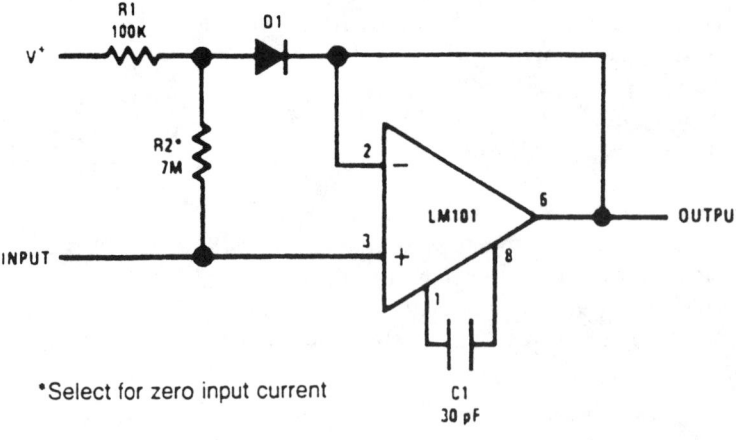

*Select for zero input current

Fig. 22. Simple bias current compensation using diode bootstrapped to the amplifier output.— Courtesy of National Semiconductor Corporation, Santa Clara, California.

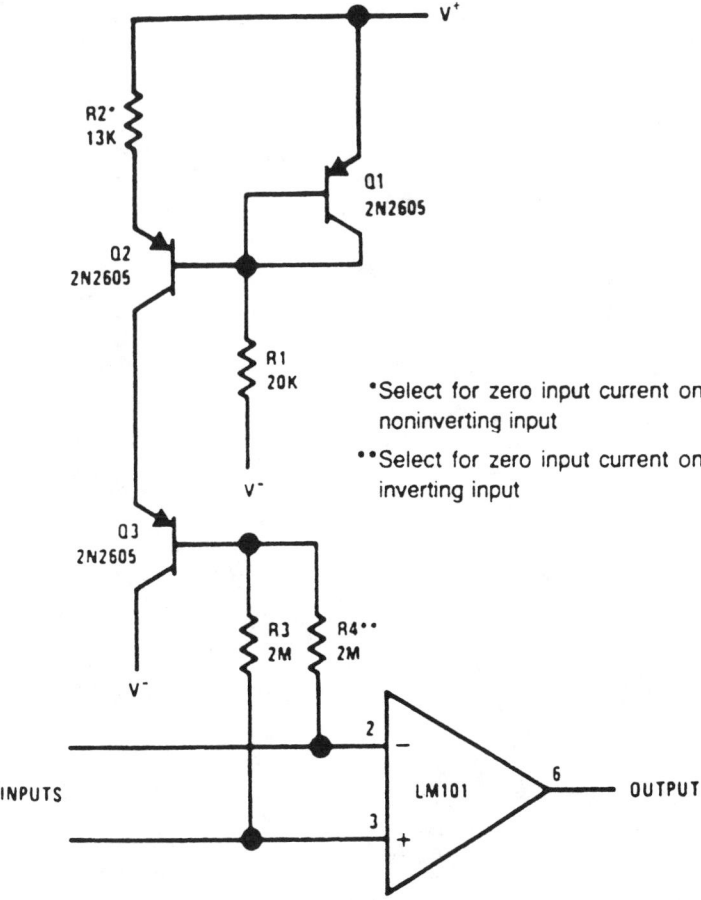

Fig. 23. Bias current compensation for differential inputs.—Courtesy of National Semiconductor Corporation, Santa Clara, California.

*Select for zero input current on noninverting input

**Select for zero input current on inverting input

Summary

This chapter has briefly touched on the IC operational amplifier, discussing its basic configurations, parameter calculation techniques, biasing, and compensation schemes. The key to the modern IC op amp is its versatility, its ability to be easily stabilized, and its low cost. Many of the potential oscillation problems encountered in amplifiers constructed from discrete components are nonexistent (or at least greatly diminished) when using op amps. Due to the combination of versatility, ease of use, and low cost, operational amplifiers are now incorporated in circuits where they were never even considered a decade ago. And such circuits are now more efficient, more useful, and/or as economical as were their forerunners that did not utilize the IC op amp. This, then, is state-of-the-art.

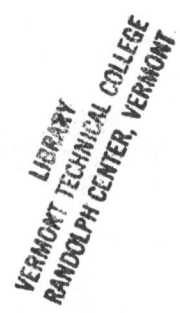

Chapter 2

Amplifier Circuits

This chapter encompasses a number of different general amplifier circuits that range from audio frequencies up into the radio-frequency spectrum. While it can be argued that any circuit which uses an op amp is in some way an amplifier, the circuits in this section may be thought of as "standard" amplifiers, in that they generally accept a low-level input and boost it to a higher level as the main function of their design.

Operational amplifiers are by and large small signal devices, but discrete components may be added to allow power output levels in the moderate to high power range. However, most op amp amplifier circuits will fall into the small signal category, where these ICs are especially useful as meter amplifiers, precision amplifiers, and, with specially chosen op amp designs, video amplifiers.

Most of these circuits lend themselves well to vectorboard construction, although it is highly desirable that when finished and tested the device is "secured" via the application of a doping compound on the leads of each component and the component body proper. Naturally, printed circuit board construction offers better stability and is preferable for certain precision circuits which address the amplification of measurement potentials and where accuracy is of paramount importance.

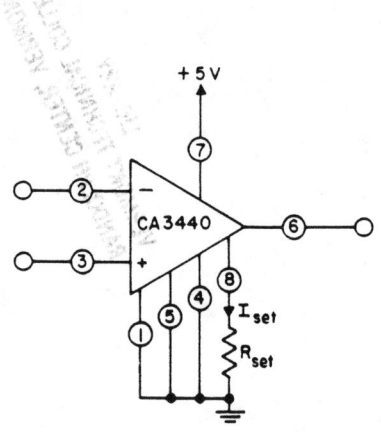

Nanopower Op Amp. This circuit is supply current programmable using R_{set} and typically requires 1 pA of input bias current over a supply voltage range of 4.0 to 15 V DC. The extremely low current demands allow this circuit to serve in applications where battery power is the sole source of power and untended operation must be maintained for long periods of time. This circuit certainly has uses in photovoltaic driven applications.— "RCA Solid State Databook— Integrated Circuits for Linear Applications." Courtesy of Harris Semiconductor.

R_{SET}	Standby Power	BW	SR
1 MΩ	250 μW	164 kHz	0.17 V/μs
10 MΩ	25 μW	27 kHz	0.017 V/μs
100 MΩ	2.5 μW	2.6 kHz	.0017 V/μs
1000 MΩ	250 nW	78 Hz	0.00017 V/μs

Single-Cell Inverting Amplifier. This simple inverting amplifier is powered from a single AA cell battery and offers 20-dB (×10) gain.—"RCA Solid State Databook—Integrated Circuits for Linear Applications." Courtesy of Harris Semiconductor.

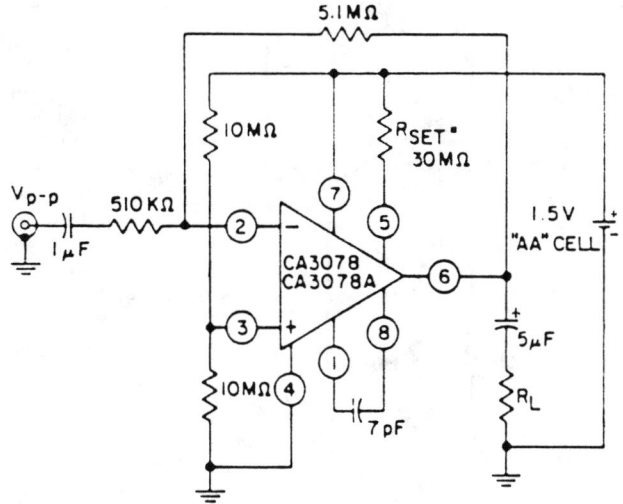

Single-Cell Noninverting Amplifier. This circuit offers a voltage gain of 10 (20 dB) and operates in the noninverting mode from a single AA cell battery. Like the preceding circuit, the simplistic power supply demands can be taken advantage of in the areas of remote operations and applications where space is at an absolute minimum.—"RCA Solid State Databook—Integrated Circuits for Linear Applications." Courtesy of Harris Semiconductor.

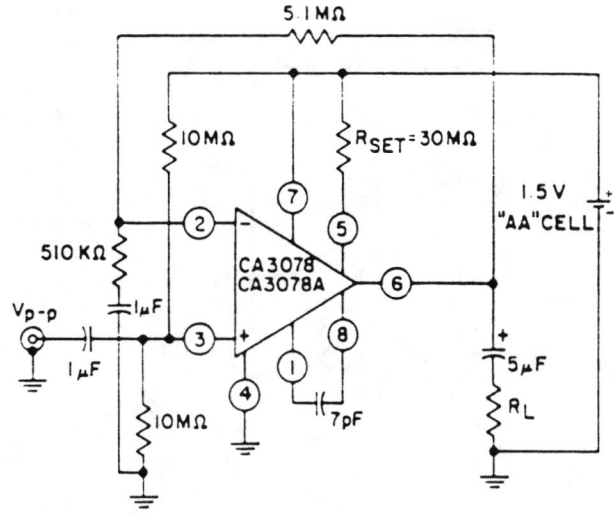

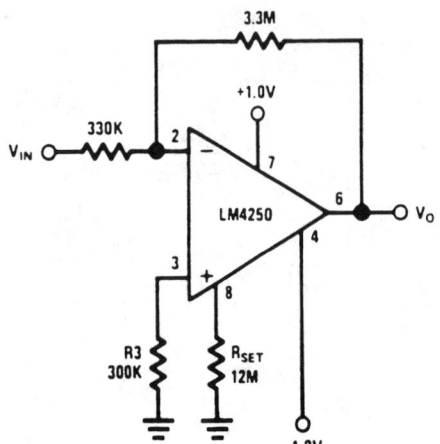

Low-Voltage, Low-Current 500-Nanowatt × 10 Amplifier.
This circuit draws a mere 260 nA from a +1-V DC power
supply. This is a bit more than the 210 nA drawn from the
−1-V DC supply due to the 12-MΩ set resistor in the
positive lead which accounts for the additional 50 nA flow.
Total power dissipation is slightly less than 500 nW.
Output voltage swing into a 100,000-Ω load is 0.7 V.
Increasing the supply voltage to 1.35 V DC increases the
output swing to a full 1.4 V peak-to-peak. This circuit
would be a good choice for applications where small
mercury cells are to be used for power. At these current
levels, normal battery shelf life may be expected.—"Linear
Applications Databook." Courtesy of National
Semiconductor Corporation, Santa Clara, California.

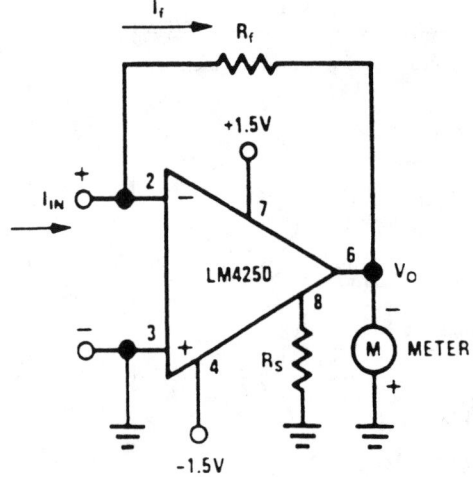

Basic Meter Amplifier.　This is a cheap and simple
circuit that operates from a 3-V DC power supply that
may be obtained from a pair of AAA batteries. The
power drain is so low when the circuit is not being
driven that an on/off switch is unnecessary. This is
a simple voltage to current converter that will output
a full scale reading of 1 μA when R_f is a 300-kΩ unit. R_s
is a 10-MΩ carbon resistor.—"Linear Applications
Databook." Courtesy of National Semiconductor
Corporation, Santa Clara, California.

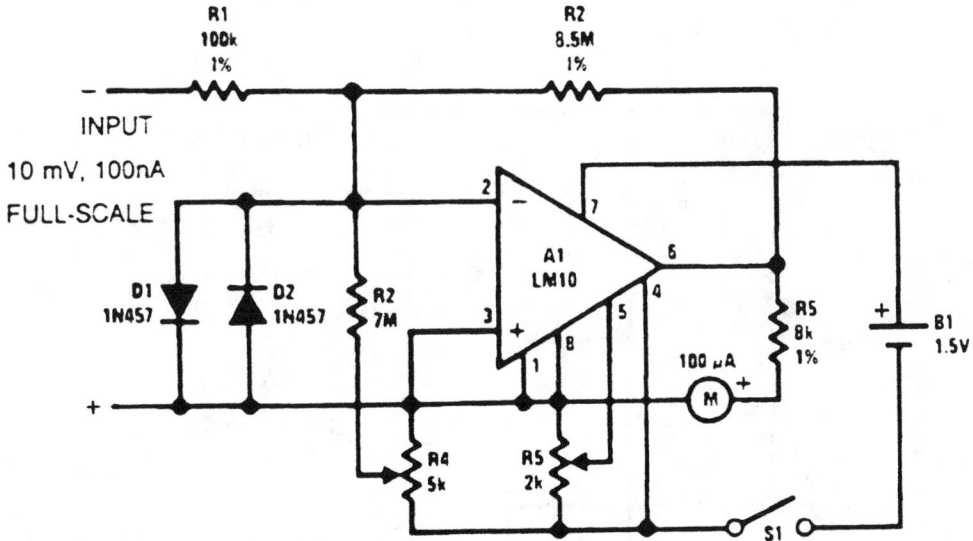

Meter Amplifier.   This is an interesting and useful circuit that offers a quick and easy meter amplifier at a minimum of cost, although component selection can be critical in certain areas. Accuracy is maintained over a 15 to 55°C range for a full scale sensitivity of 10 mV and 100 nA using this circuit, although negligible zero drift has been measured with 1 mV and 10 nA sensitivities. The meter zeroing circuit operates from the reference output, so it should not be necessary to make frequent adjustments as the battery voltage changes due to age or use. The only critical components are the two potentiometers which should be precision types. If inexpensive units are used here, it will be quite difficult to zero the meter (and keep it zeroed).—"Linear Databook I." Courtesy of National Semiconductor Corporation, Santa Clara, California.

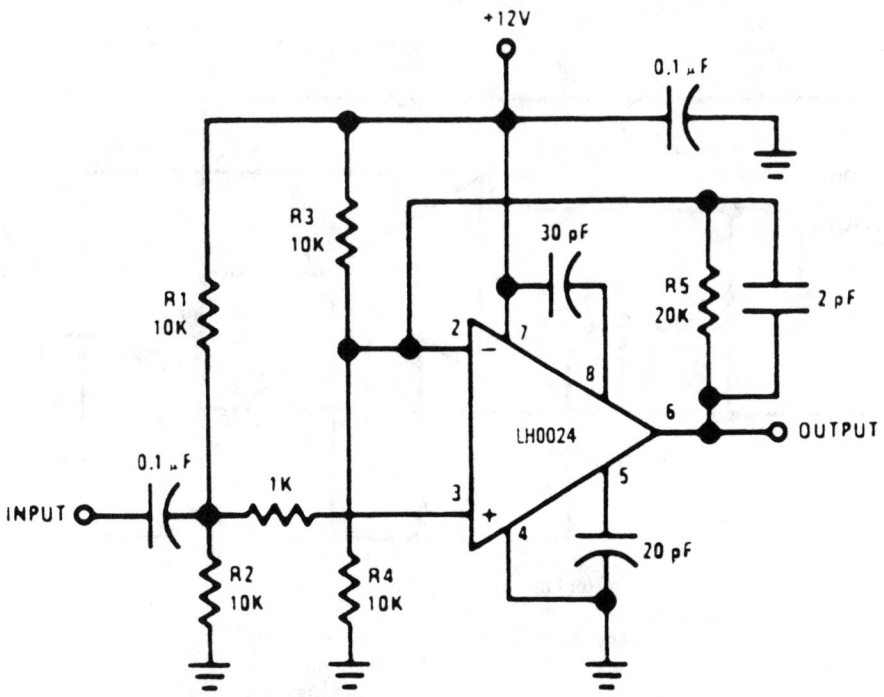

Video Amplifier.   The LH0024 is a very wide bandwidth, high slew rate op amp intended to fulfill a wide variety of high-speed applications. It exhibits a useful gain at frequencies in excess of 50 MHz, making it practical for use in video applications. This circuit is simple and uses a single polarity power supply and thus less stress is put on adding it to already existing circuitry. Supply voltage can vary but should not exceed 18 V DC.—"Linear Databook I." Courtesy of National Semiconductor Corporation, Santa Clara, California.

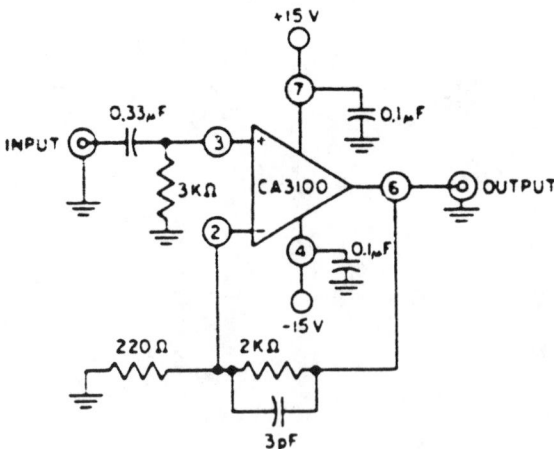

20-Decibel Video Amplifier.   This amplifier exhibits a bandwidth of about 20 MHz and uses the RCA CA3100 wideband op amp. The circuit uses a minimum of components and will operate satisfactorily from a dual polarity 12-V supply with slightly diminished bandwidth.—"RCA Solid State Databook— Integrated Circuits for Linear Applications." Courtesy of Harris Semiconductor.

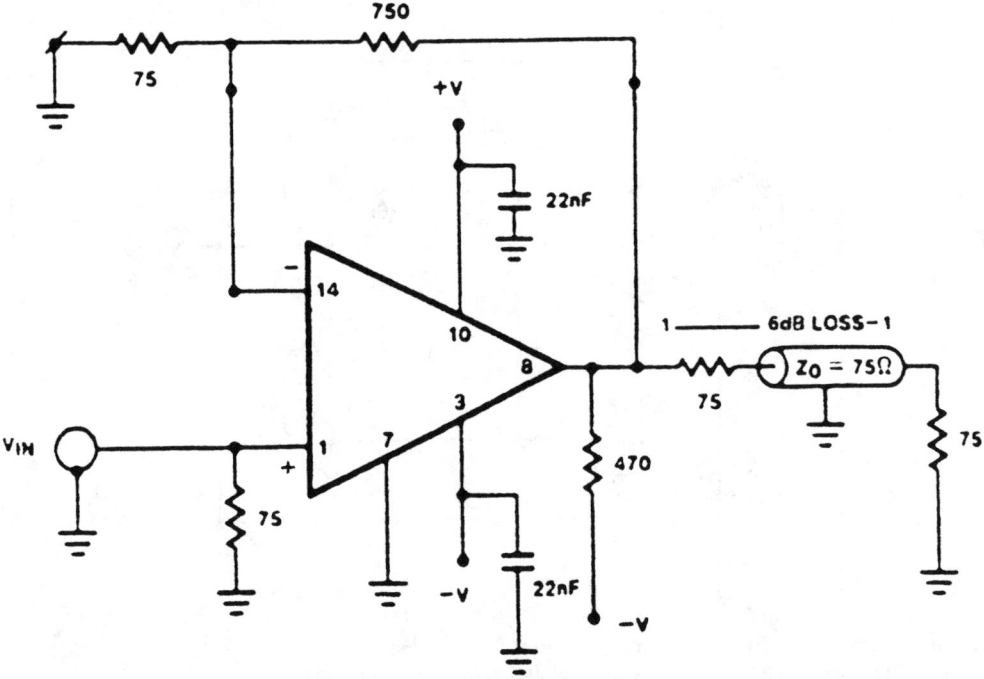

Video Amplifier. This color video amplifier uses the Signetics NE5539 UHF op amp in a circuit that offers less than 0.5% gain variation from staircase top to bottom.—"Linear Data Manual Volume 2: Industrial." Courtesy of Signetics Company, a division of North American Philips Corporation.

Wideband, Low-Noise, Low-Drift Amplifier. The bandwidth exhibited by this circuit is approximately 240 kHz using a Motorola LF356 JFET input op amp or equivalent. C1 is a 3 pF unit. The compensation network consists of C2 and R2. The relationship is R2C2 = R1C1.—"Motorola Linear and Interface Integrated Circuits." Courtesy of Motorola Semiconductor Products, Motorola, Inc.

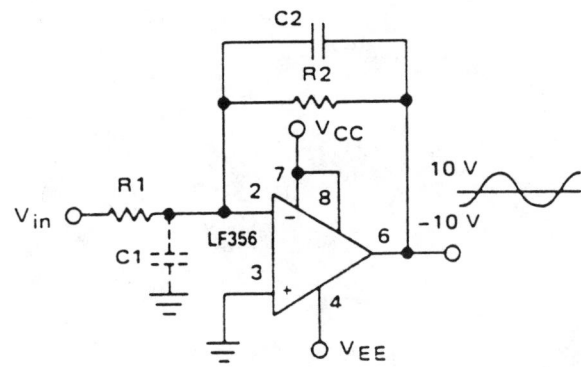

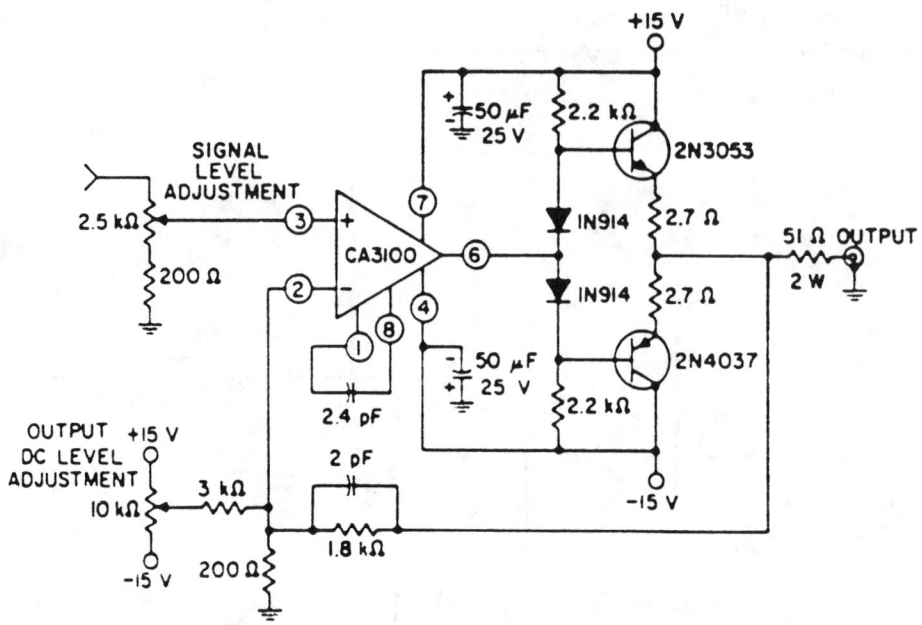

2-Watt Wideband Amplifier. This high slew rate, wideband amplifier uses a bipolar transistor output stage to supply 2 W to a 50-Ω load. Its nominal bandwidth is 10 MHz.—"RCA Solid State Databook—Integrated Circuits for Linear Applications." Courtesy of Harris Semiconductor.

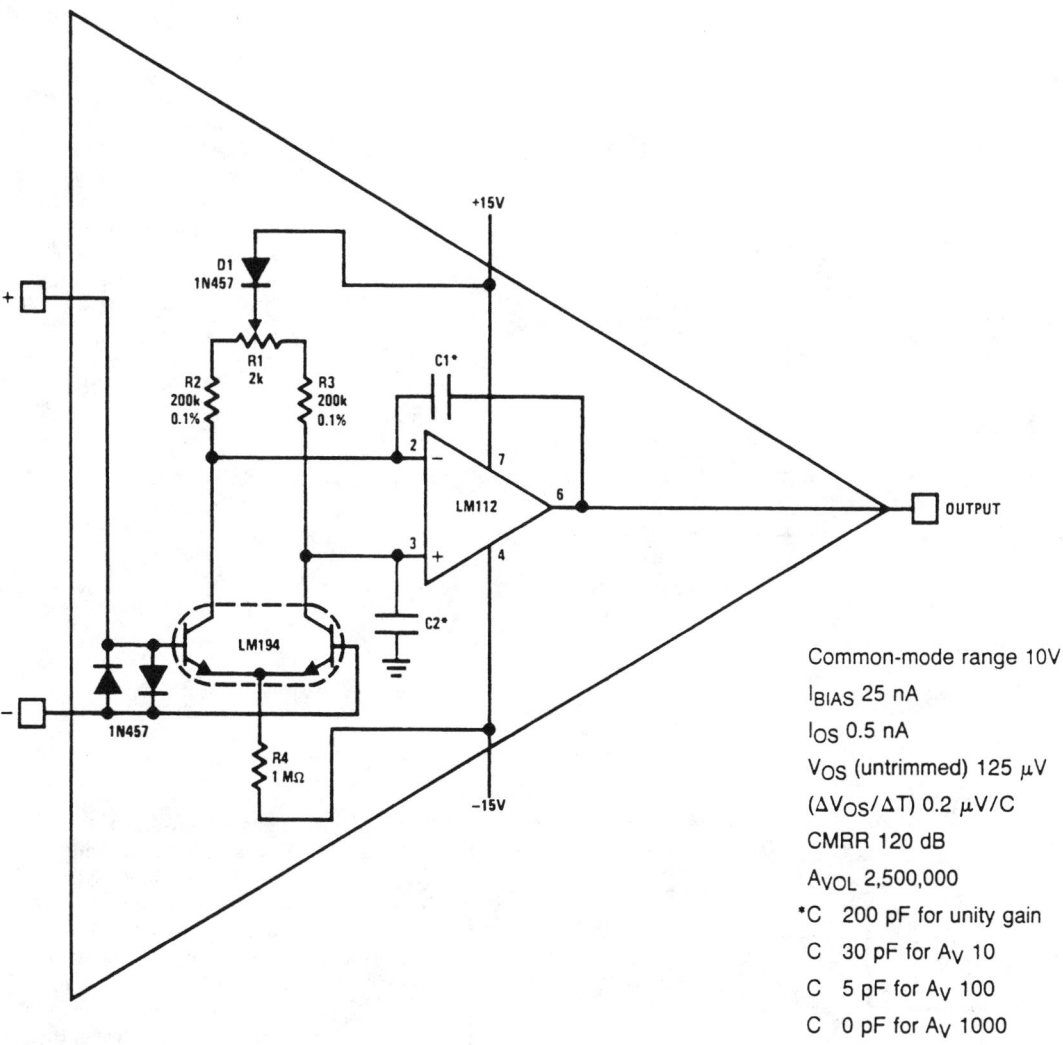

Common-mode range 10V

I_{BIAS} 25 nA

I_{OS} 0.5 nA

V_{OS} (untrimmed) 125 μV

($\Delta V_{OS}/\Delta T$) 0.2 μV/C

CMRR 120 dB

A_{VOL} 2,500,000

*C   200 pF for unity gain

C   30 pF for A_V 10

C   5 pF for A_V 100

C   0 pF for A_V 1000

Precision Low-Drift Operational Amplifier.   This circuit combines the LM112 common op amp with the LM194 Supermatch Pair, a junction-isolated well-matched monolithic N-P-N transistor pair, to form a highly precise low-drift op amp. Gain is adjusted by varying the value of C2. 200 pF yields unity gain. Decreasing the capacitance to 30 pF results in a gain of 10.—"Linear Databook III." Courtesy of National Semiconductor Corporation, Santa Clara, California.

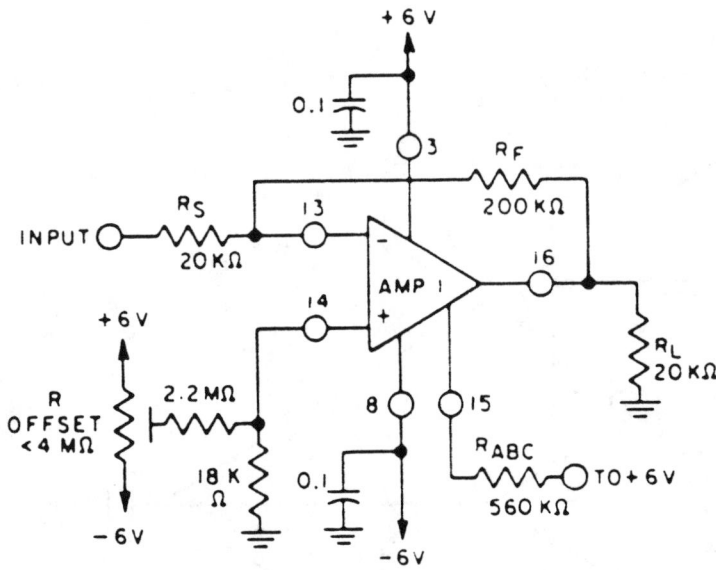

20-Decibel Amplifier. This amplifier offers closed-loop voltage gain of 10 for 20 dB and works from a bipolar 6-V DC supply. Maximum input voltage is 50 mV, while the input/output resistance is 20 kΩ. The op amp is an RCA CA3060 (of which only one of the three op amps is used) or equivalent.—"RCA Solid State Databook—Integrated Circuits for Linear Applications." Courtesy of Harris Semiconductor.

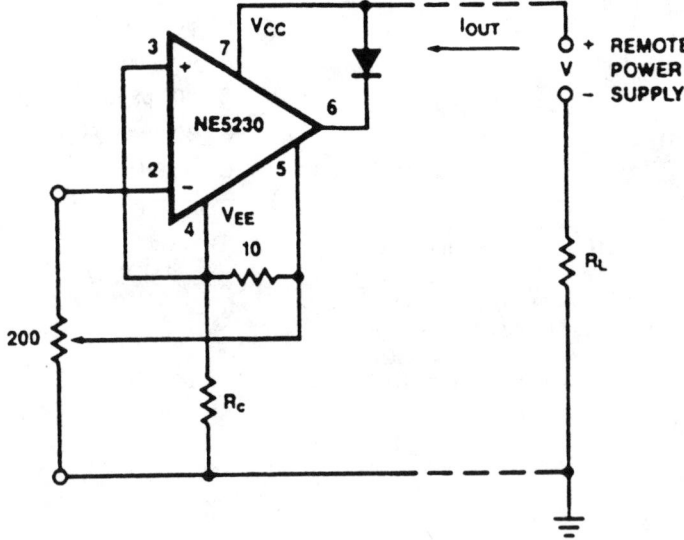

Remote Transducer Transconductance Amplifier. This circuit uses the NE5230 in a pseudo-transistor configuration with the inverting input equivalent to the transistor base. The point where the noninverting input and VEE meet is the emitter, and the junction of the diode at pin 7 is the collector. This is a more or less generic circuit that accepts the input from one of many different types of transducers. It produces a current transmission output of 4 to 20 mA in direct relationship with the transducer input. The diode is essential in order to prevent saturation.—"Linear Data Manual Volume 2: Industrial." Courtesy of Signetics Company, a division of North American Philips Corporation.

Absolute Value Amplifier.
This circuit generates a positive output voltage for either polarity of input. For positive signal inputs, it acts as a noninverting amplifier, and it acts as an inverting amplifier for negative inputs. Keep input values above 1 V as accuracy is poor below this level.—"Linear Data Manual Volume 2: Industrial." Courtesy of Signetics Company, a division of North American Philips Corporation.

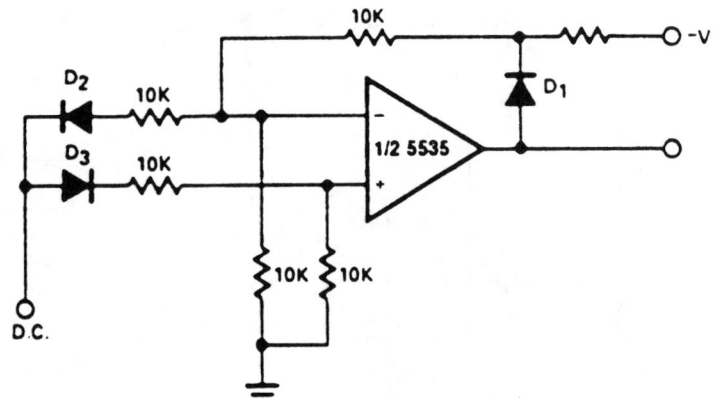

Summing Amplifier. This summing amplifier exhibits minimal input current requirements and a power bandwidth of 250 kHz. (Small signal bandwidth = 3.5 MHz) In addition to increasing speed, the LM101A raises high- and low-frequency gain, increases output drive capability, and eliminates thermal feedback. C5 = 0.0000006/R1.—"Motorola Linear and Interface Integrated Circuits." Courtesy of Motorola Semiconductor Products, Motorola, Inc.

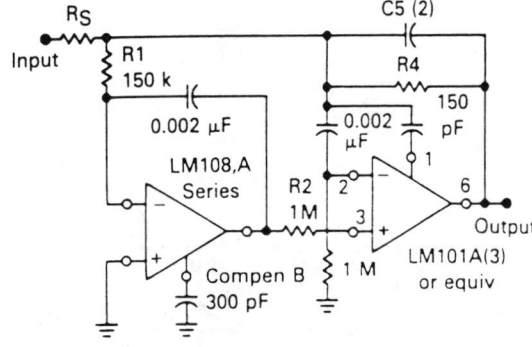

Logarithmic Amplifier. The Motorola MC1456 series of internally compensated op amps is used for this circuit which also incorporates a common N-P-N transistor. The 10-kΩ potentiometer is the offset adjustment.—"Motorola Linear and Interface Integrated Circuits." Courtesy of Motorola Semiconductor Products, Motorola, Inc.

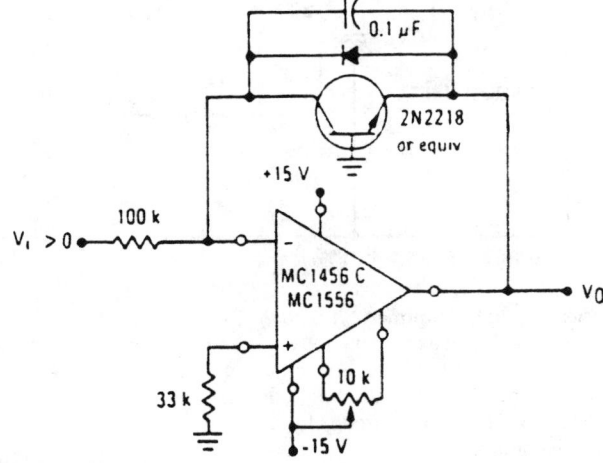

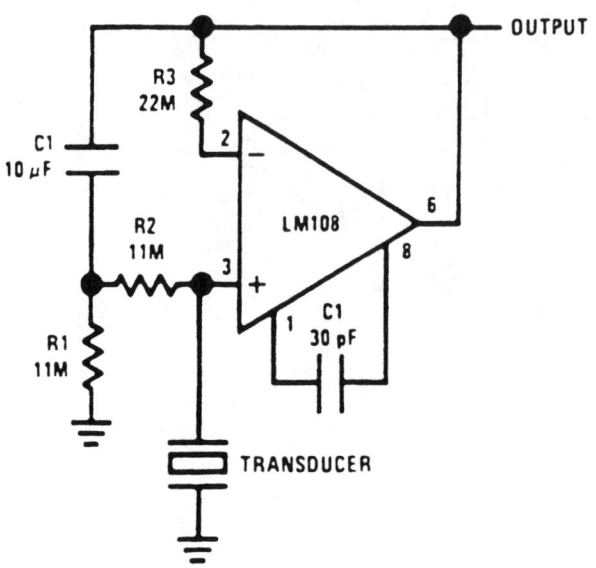

Piezoelectric Amplifier. This is a general purpose amplifier to be driven by piezoelectric transducers. Such sensors normally require a high input resistance amplifier. Using the LM108, input resistances on the order of 100 MΩ maximum can be achieved.—"Linear Applications Databook." Courtesy of National Semiconductor Corporation, Santa Clara, California.

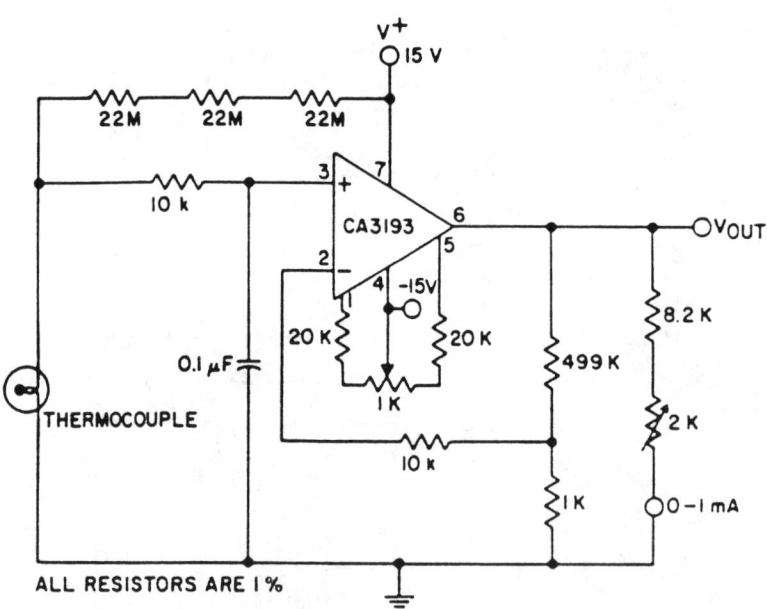

Thermocouple Amplifier. The RCA CA3193 is an ultrastable instrumentation op amp that is an excellent choice for thermocouple amplification. This circuit amplifies the output from the thermocouple 500 times. The three 22-MΩ resistors provide full scale should the thermocouple open.—RCA Solid State Databook—Integrated Circuits for Linear Applications." Courtesy of Harris Semiconductor.

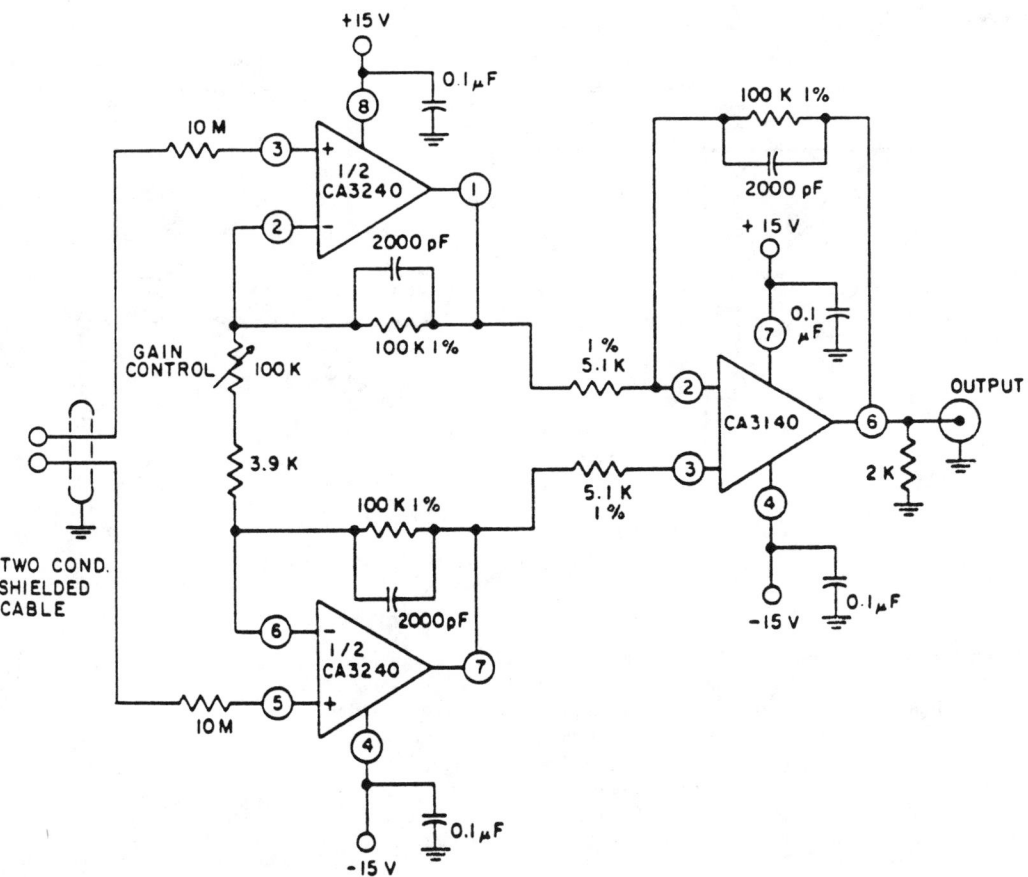

Precision Differential Amplifier.   This amplifier offers extremely high input impedance, making it desirable for biomedical monitoring applications where electrodes are connected directly to the human body. Here, the 10-MΩ series resistors limit any current that might result in patient discomfort. Current is held to about 2 μA.—"RCA Solid State Databook—Integrated Circuits for Linear Applications." Courtesy of Harris Semiconductor.

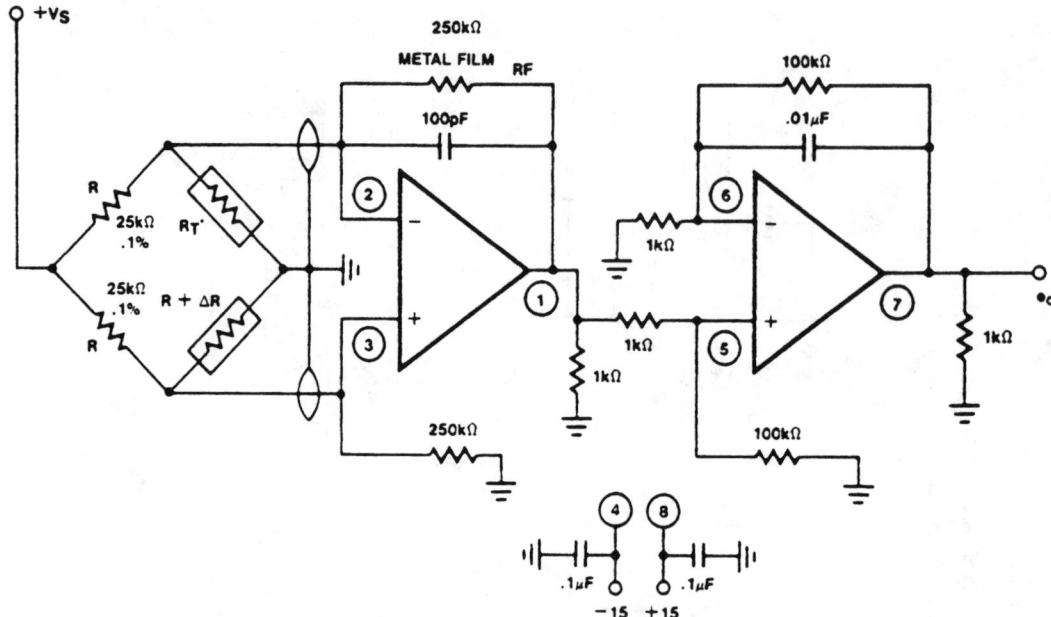

Bridge Transducer Amplifier. A bridge amplifier is used in applications involving strain gauges, thermal sensors, and accelerometers. Most sensor units are high-impedance types requiring an amplifier with an equally high-resistance input for best sensitivity. This amplifier has a high-impedance input, low bias current, and low drift characteristics. The op amp is a Signetics SE5512 or equivalent and provides a gain of 50 in the first stage and 100 in the second.—"Signetics Linear Data Manual Volume 2: Industrial." Courtesy of Signetics Company, a division of North American Philips Corporation.

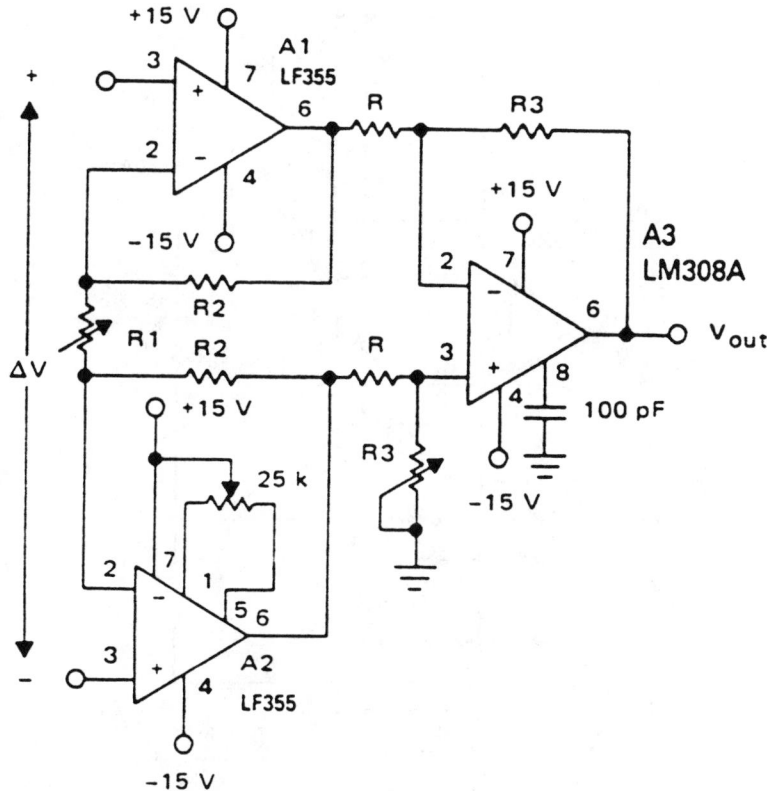

High-Impedance, Low-Drift Instrumentation Amplifier.   This circuit uses three ICs and a 15-V DC bipolar supply to produce an output equal to R3/ R(2R2/R1 + 1). R3 is adjusted for 120-dB boost.—"Motorola Linear and Interface Integrated Circuits." Courtesy of Motorola Semiconductor Products, Motorola, Inc.

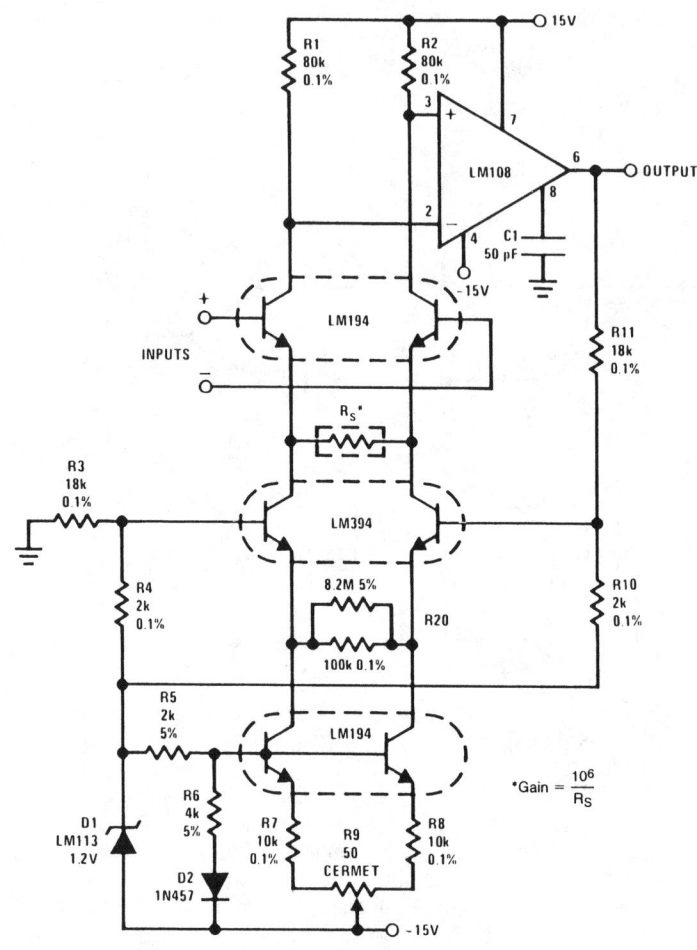

**High-Performance
Instrumentation Amplifier.**
This instrumentation
amplifier circuit uses the
LM108 in combination with
the LM194 and LM394
Supermatch Pair monolithic
N-P-N transistor pairs. Gains
of from 10 to 10,000 are easily
achieved.—"Linear
Databook III." Courtesy of
National Semiconductor
Corporation, Santa Clara,
California.

$$\text{*Gain} = \frac{10^6}{R_S}$$

Performance Characteristics

	G = 10,000	G = 1,000	G = 100	G = 10	
Linearity of Gain (± 10V Output)	≤0.01	≤0.01	≤0.02	≤0.05	%
Common-Mode Rejection Ratio (60 Hz)	≥ 120	≥ 120	≥ 110	≥ 90	dB
Common-Mode Rejection Ratio (1 kHz)	≥ 110	≥ 110	≥ 90	≥ 70	dB
Power Supply Rejection Ratio					
+ Supply	> 110	> 110	> 110	> 110	dB
− Supply	> 110	> 110	> 90	> 70	dB
Bandwidth (− 3 dB)	50	50	50	50	kHz
Slew Rate	0.3	0.3	0.3	0.3	V/μs
Offset Voltage Drift**	≤0.25	≤0.4	2	≤ 10	μV/°C
Common-Mode Input Resistance	> 10^9	> 10^9	> 10^9	> 10^9	Ω
Differential Input Resistance	> 3 x 10^8	> 3 x 10^8	> 3 x 10^8	> 3 x 10^8	Ω
Input Referred Noise (100 Hz ≤ f ≤ 10 kHz)	5	6	12	70	$\frac{nV}{\sqrt{Hz}}$
Input Bias Current	75	75	75	75	nA
Input Offset Current	1.5	1.5	1.5	1.5	nA
Common-Mode Range	± 11	± 11	± 11	± 10	V
Output Swing (R_L = 10 kΩ)	± 13	± 13	± 13	± 13	V

**Assumes ≤ 5 ppm/°C tracking of resistors

Amplifier and Line Driver.
The output from the op amp
is coupled to a push/pull
transistor amplifier and is
designed to drive a 50-Ω line.
The op amp is a Motorola
MC3301 single-supply
operating from a 15-V DC
unipolar supply.—"Motorola
Linear and Interface
Integrated Circuits."
Courtesy of Motorola
Semiconductor Products,
Motorola, Inc.

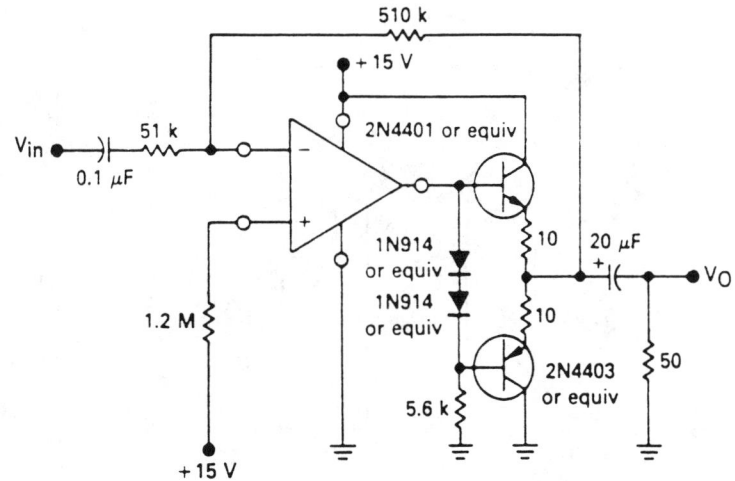

Variable Gain Circuit. The 5320 low-
voltage op amp is configured in a
noninverting gain of five. The output drives
the SD210 DMOS FET whose series resistance
changes with this output voltage, which in
turn changes the gain of the NE5212
transimpedance amplifier.—"Signetics Linear
Data Manual Volume 1: Communications."
Courtesy of Signetics Company, a division of
North American Philips Corporation.

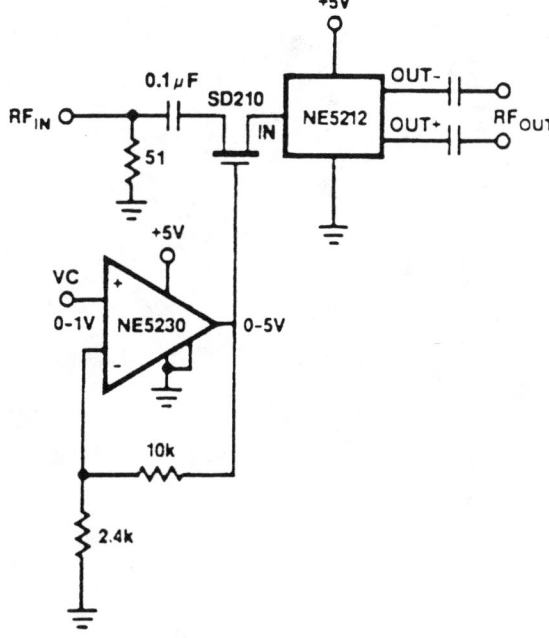

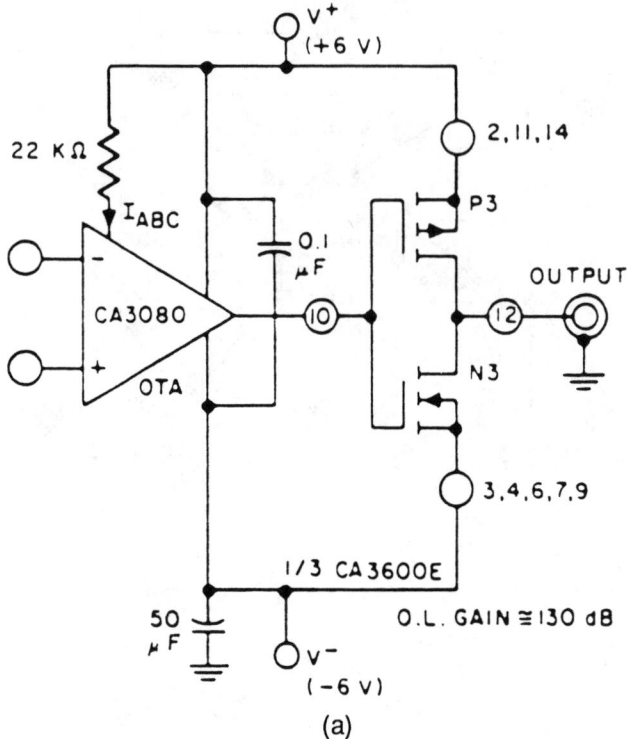

(a)

Post-Amplifier for Op Amps (*four circuits*).   CMOS transistor pairs can be effectively used as post-amplifiers for op amps. Due to the high input resistance typically exhibited by a CMOS pair, the op amp operates under essentially unloaded conditions. (*a*) The first circuit couples the output of the op amp to the RCA CA3600 CMOS pair. The 30-dB gain in a single CMOS pair is an added increment to the 100-dB gain in the CA3080. This results in a total forward output of approximately 130 dB. This is an open-loop circuit.— "RCA Solid State Databook— Integrated Circuits for Linear Applications." Courtesy of Harris Semiconductor.

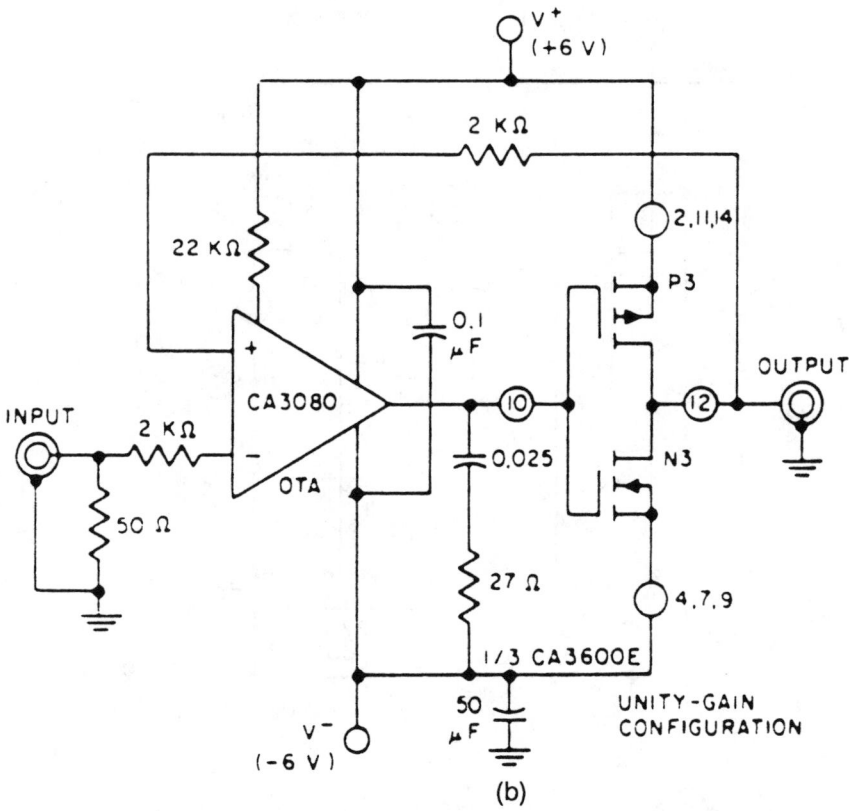

Post-Amplifier for Op Amps (b).   The second schematic depicts a post-amplifier configuration in a unity gain circuit. The open-loop slew rate is normally about 65 V/μsec, but when compensated for the unity gain voltage-follower mode, the rate is adjusted to approximately 1 V/μsec.—"RCA Solid State Databook—Integrated Circuits for Linear Applications." Courtesy of Harris Semiconductor.

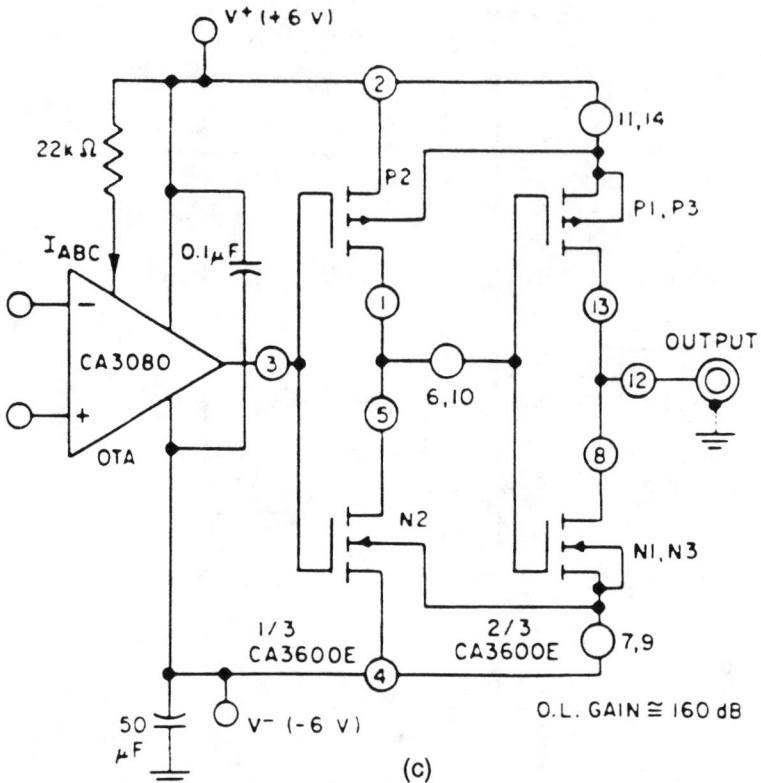

Post-Amplifier for Op Amps (c).   The use of two-stage CMOS post-amplifier configurations as represented by the third circuit increases the total open-loop gain of the system to about 160 dB with an open-loop slew rate still at the 65 V/μsec rate.—"RCA Solid State Databook—Integrated Circuits for Linear Applications." Courtesy of Harris Semiconductor.

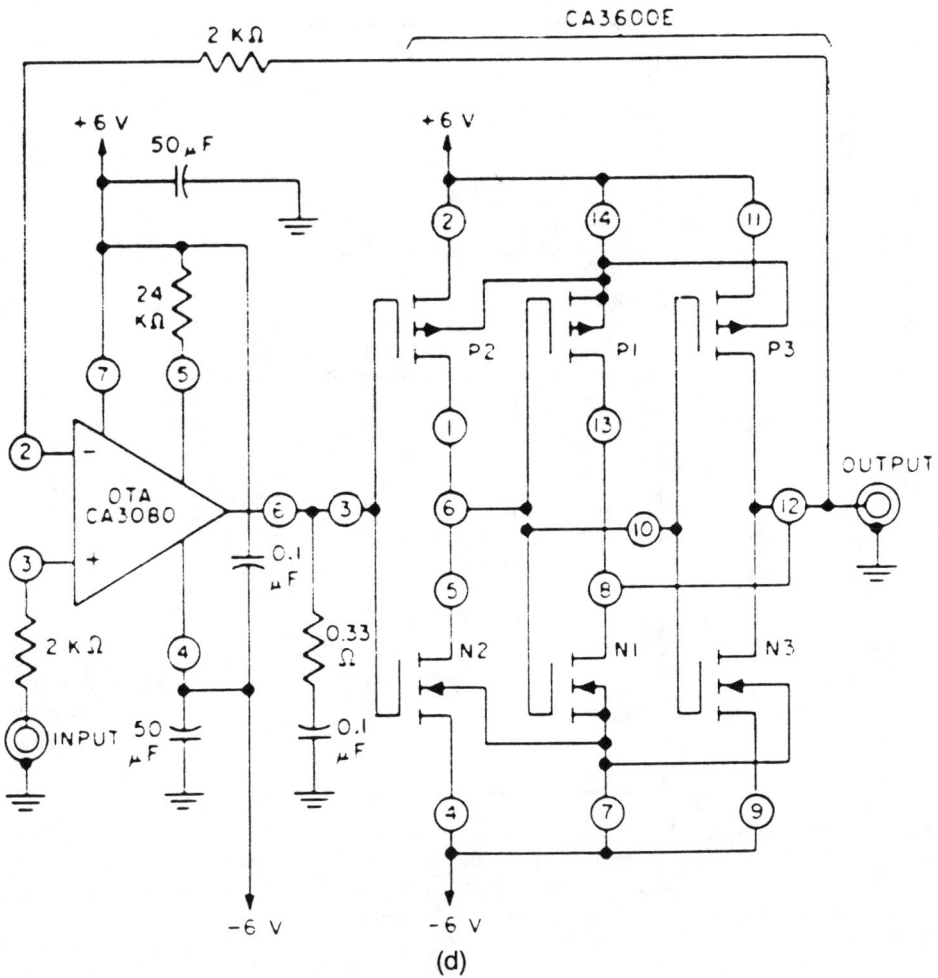

Post-Amplifier for Op Amps (d).   The fourth circuit connected in the unity gain follower mode maintains a slew rate of 1 V/μsec.—"RCA Solid State Databook—Integrated Circuits for Linear Applications." Courtesy of Harris Semiconductor.

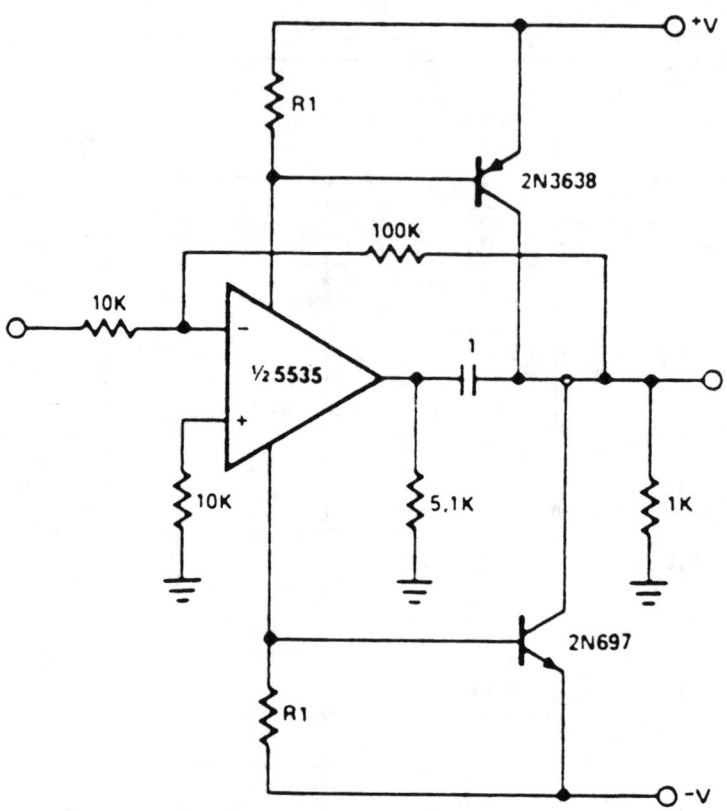

Op Amp Power Booster. Usually, the output power from most op amps is sufficient to drive other devices with relatively low input signal requirements; however, the available output may occasionally be insufficient due to design problems, higher than anticipated circuit losses, or electronic stages that require more drive than anticipated. If the lack of power is not too severe, this circuit may help avoid the possibility of a complete redesign of a complex circuit. This is a simple power booster designed to accept the input from a typical op amp and boost it to a level that will allow moderate loads to be driven. The boost is derived from the N-P-N transistor and adds little to circuit complexity, design criteria, or cost. This circuit might save an otherwise botched circuit design.—"Linear Data Manual Volume 2: Industrial." Courtesy of Signetics Company, a division of North American Philips Corporation.

Current Amplifier. Low current is supplied at the input potential as the power supply to load resistor R_L. The load current is increased by the multiplication factor of R2/R1 when the load current is monitored by meter M. Therefore, if the load current is 100 nA, the current drain from the power supply is 100 μA, a value more easily measured. The input and output voltages are transferred at the same potential and only the output current is multiplied by the scale factor.— "RCA Solid State Databook— Integrated Circuits for Linear Applications." Courtesy of Harris Semiconductor.

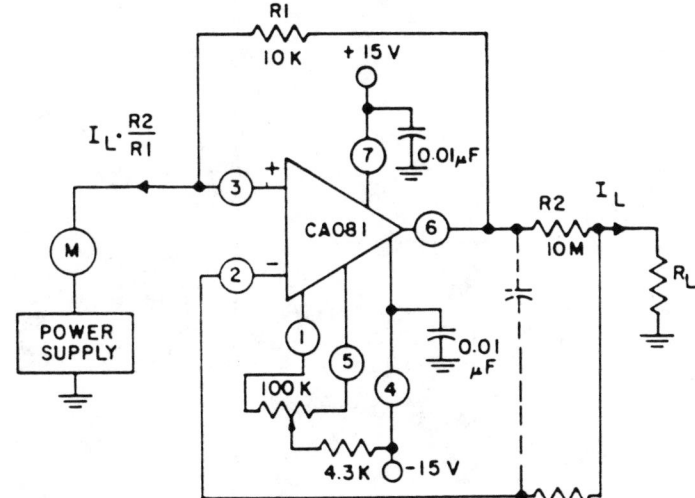

Chapter 3

Oscillators/Multivibrators

Operational amplifiers are often utilized in oscillator circuits. However, their capabilities are usually quite diminished when the standard audio frequency spectrum is exceeded. Therefore, the circuits in this chapter generally are designed to operate within or (sometimes) just above the audio frequency range. As such, they lend themselves well to siren/alert type applications and other circuits that can generate pulses with durations of a few microseconds to a second or more.

Many of these circuits form the building blocks for more elaborate test instrumentation, the circuits for which are featured in Chapter 5 of this text. Through the application of proper filtering techniques, the output from op amp oscillators and multivibrators can be altered to produce very clean sine waves or other wave shapes of a precise nature. All of this is obtained with fairly simplistic circuits using the op amp as the core device. It is a relatively easy task from a design and construction standpoint to build a circuit that offers great mechanical stability, owing to the small space requirements of the single op amp package and the few discrete components needed. In many instances, such circuits are further stabilized mechanically by "potting" them in epoxy or other similar compounds.

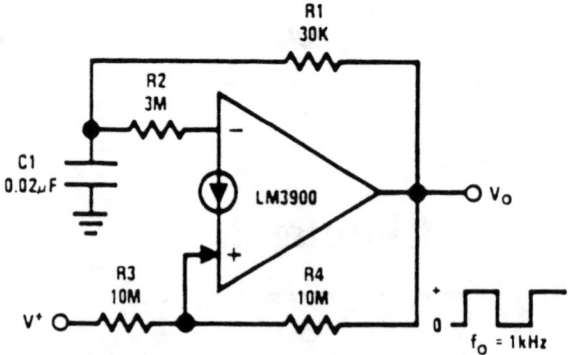

Square Wave Oscillator. This circuit is about as simple, cheap, and easy to construct as possible. The key to operation is the single capacitor that charges and discharges between the voltage limits established by R2, R3, and R4. The charging is accomplished through the 30-kΩ resistor between the output and input. This combination produces a Schmitt Trigger circuit. When the output is low, the trigger will be fired via R2 as its current approaches and equals the current at the positive input. Design output frequency for this circuit using the component values shown is 1000 Hz.— "Linear Databook I." Courtesy of National Semiconductor Corporation, Santa Clara, California.

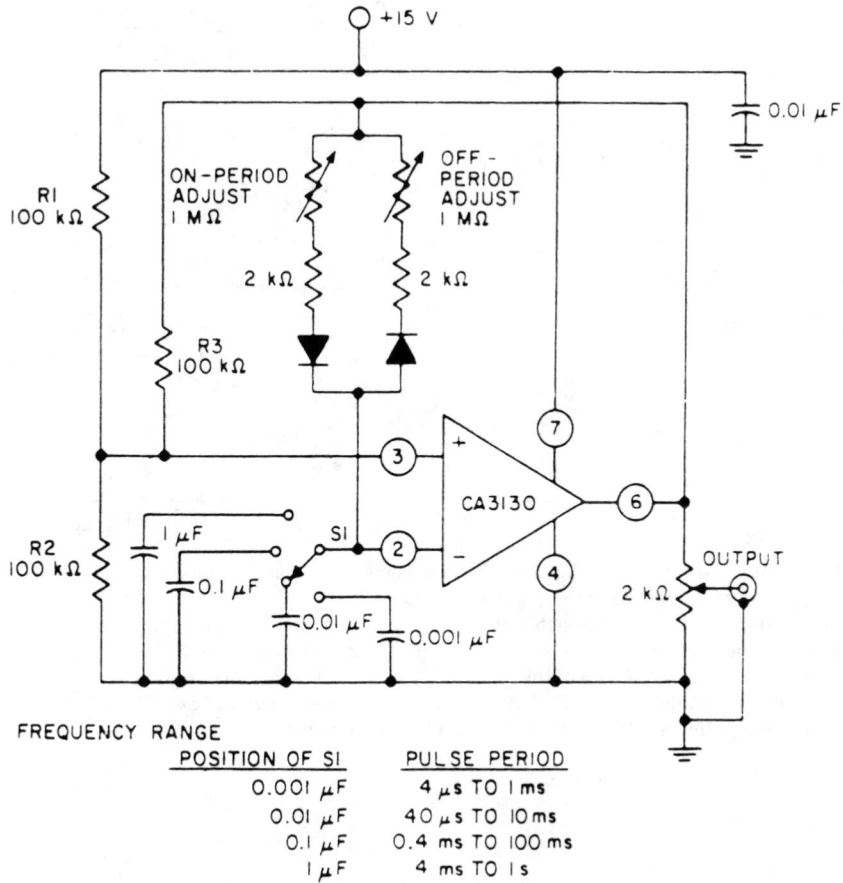

FREQUENCY RANGE

POSITION OF SI	PULSE PERIOD
0.001 μF	4 μs TO 1 ms
0.01 μF	40 μs TO 10 ms
0.1 μF	0.4 ms TO 100 ms
1 μF	4 ms TO 1 s

Astable Multivibrator. This circuit generates pulses over a period range of from 4 μsec to 1 sec, depending on the position of the frequency range switch. R1 and R2 bias the op amp to the midpoint of the supply voltage, while R3 is the sole feedback resistor. Other period ranges are easily obtained by adjusting the values of the capacitors in the timing portion of the circuit.—"RCA Solid State Databook—Integrated Circuits for Linear Applications." Courtesy of Harris Semiconductor.

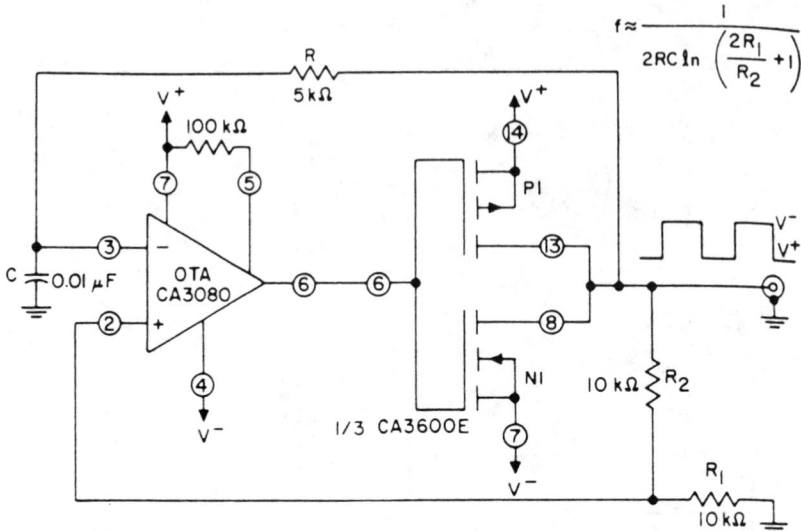

$$f \approx \frac{1}{2RC \ln \left(\dfrac{2R_1}{R_2} + 1 \right)}$$

Astable Multivibrator. A CA3600E CMOS array is used in conjunction with an operational transconductance amplifier to form this circuit. Precise timing thresholds are assured by the stable characteristics of the differential amplifier. Speed versus power consumption tradeoffs can be made by adjusting the bias current applied to terminal 5 of the op amp. The resting power consumption of this circuit is about 6 mW.—"RCA Solid State Databook—Integrated Circuits for Linear Applications." Courtesy of Harris Semiconductor.

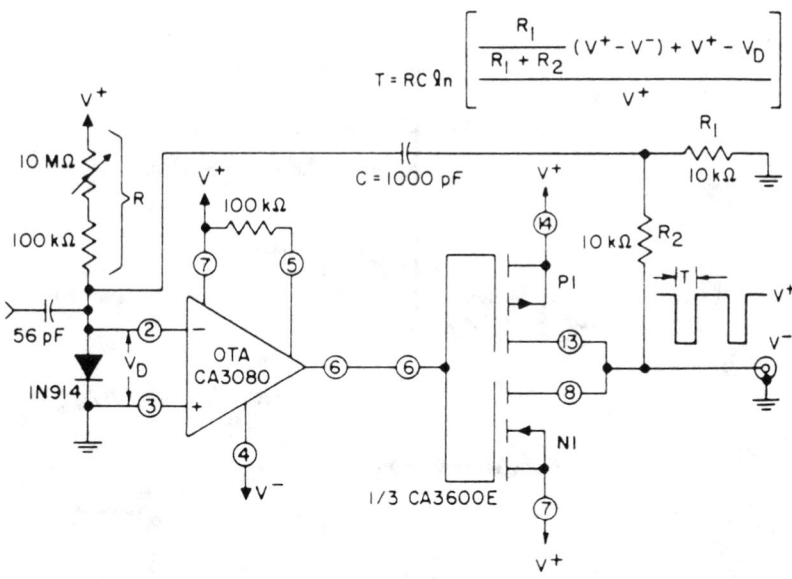

$$T = RC \ln \left[\frac{\dfrac{R_1}{R_1 + R_2}(V^+ - V^-) + V^+ - V_D}{V^+} \right]$$

Monostable Multivibrator.   This is the monostable equivalent of the previous circuit. The same descriptions and ratings apply.—"RCA Solid State Databook—Integrated Circuits for Linear Applications." Courtesy of Harris Semiconductor.

Sine Wave Generator.   This circuit uses a Wein Bridge configuration. It is stable because it is a simple design that requires few components, thus the mechanical aspect of construction is far less complex. It uses both positive and negative feedback and oscillation will cease if too much of either is used.— "Linear Data Manual Volume 2: Industrial." Courtesy of Signetics Company, a division of North American Philips Corporation.

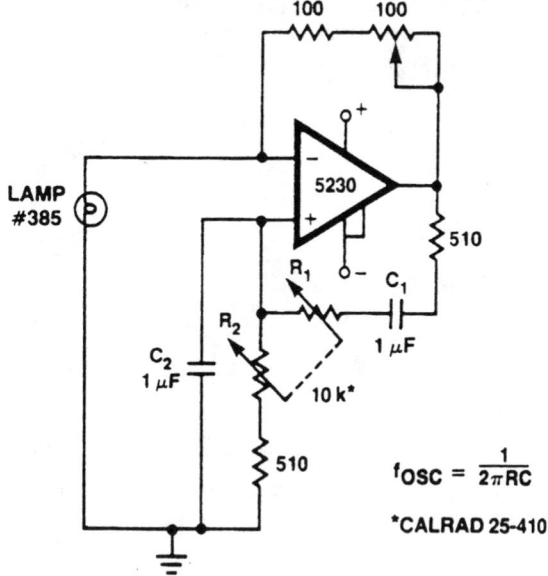

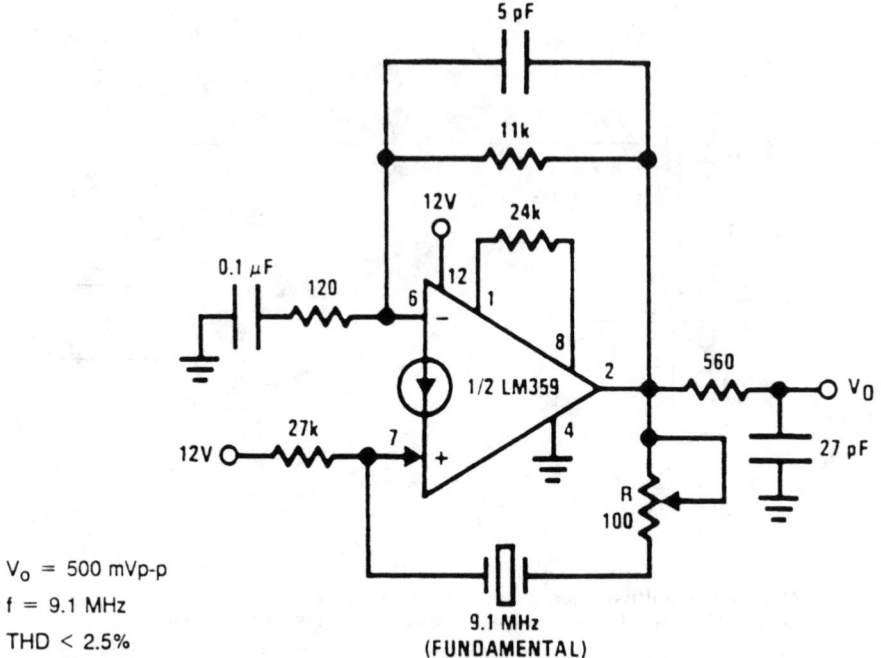

V_0 = 500 mVp-p

f = 9.1 MHz

THD < 2.5%

Crystal-Controlled Sine Wave Oscillator.   This design uses the superfast current differencing Norton amplifier in a circuit that has an output frequency at the fundamental frequency of the crystal. Output is rated at 500 mV peak-to-peak and the construction is extremely simple and straightforward, making it easy to obtain good mechanical rigidity for enhanced frequency stability. Additionally, the circuit operates from a single polarity 12-V DC supply, making it highly practical to incorporate this circuit in an existing piece of equipment.—"Linear Databook I." Courtesy of National Semiconductor Corporation, Santa Clara, California.

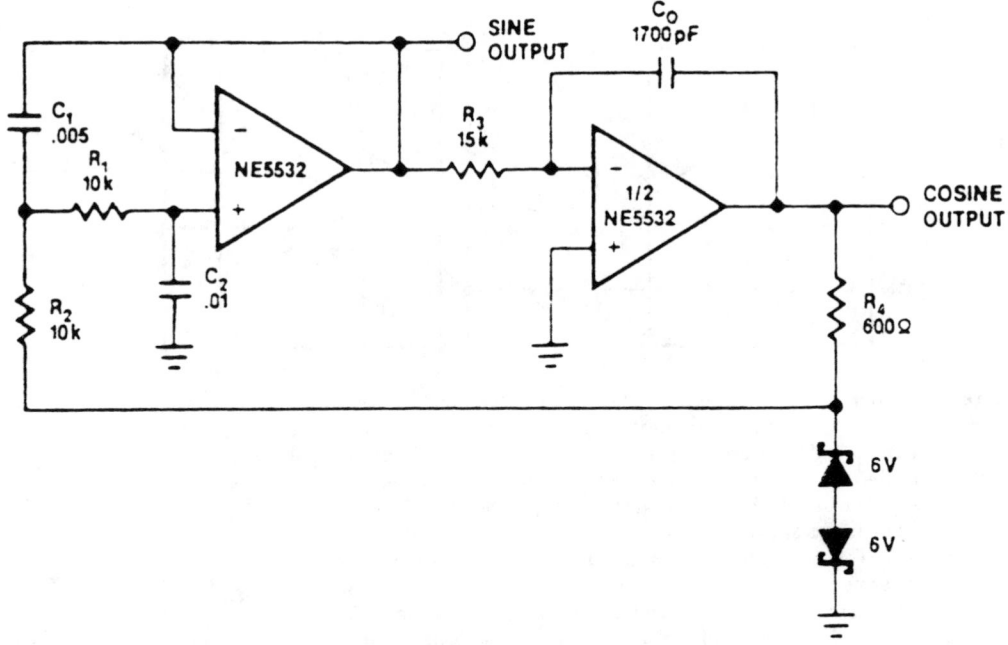

Two-Phase Sine Wave Oscillator. A two-pole Butterworth filter followed by a phase shifter single-pole stage allows this circuit to output a two-phase sine wave (sine/cosine). Using the components shown, the output frequency will be approximately 2000 Hz. Other RC values can easily be substituted for other output frequencies. This circuit was designed around a conventional 741 op amp, but the Signetics device should offer better distortion characteristics, especially at higher frequencies.—"Linear Data Manual Volume 2: Industrial." Courtesy of Signetics Company, a division of North American Philips Corporation.

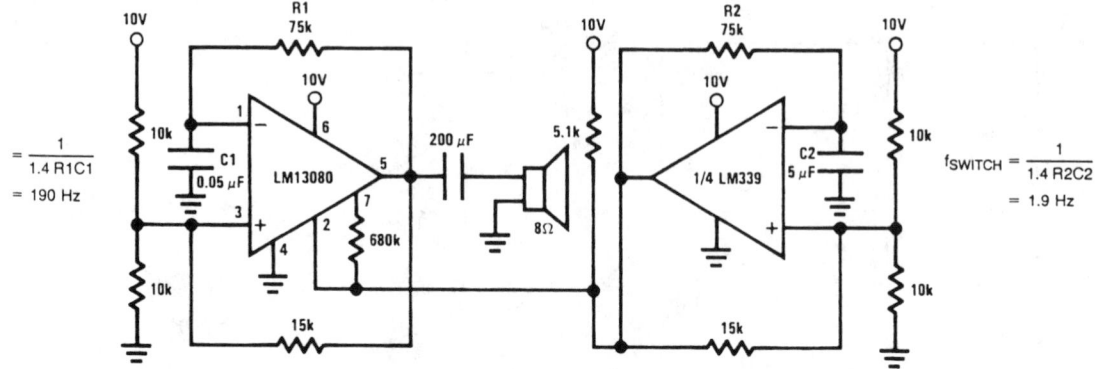

Two-State Siren. This circuit uses the LM13080 Programmable op amp and one quarter of the LM339 quad comparator to output two separate tones. In this configuration the LM339 serves as a switch in order to go to the second tonal state. Using the components designated here, the low tone (1st condition) will be about 200 Hz while the high (second state) tone is about 2 kHz. The output is sufficient to drive a small 8-Ω speaker. A 9-V battery may serve as a portable power source. This is an interesting circuit, but there are far simpler ways to arrive at the same type of output using single ICs or even a single transistor. It is presented here for those persons who like to do things the hard way.— "Linear Databook I." Courtesy of National Semiconductor Corporation, Santa Clara, California.

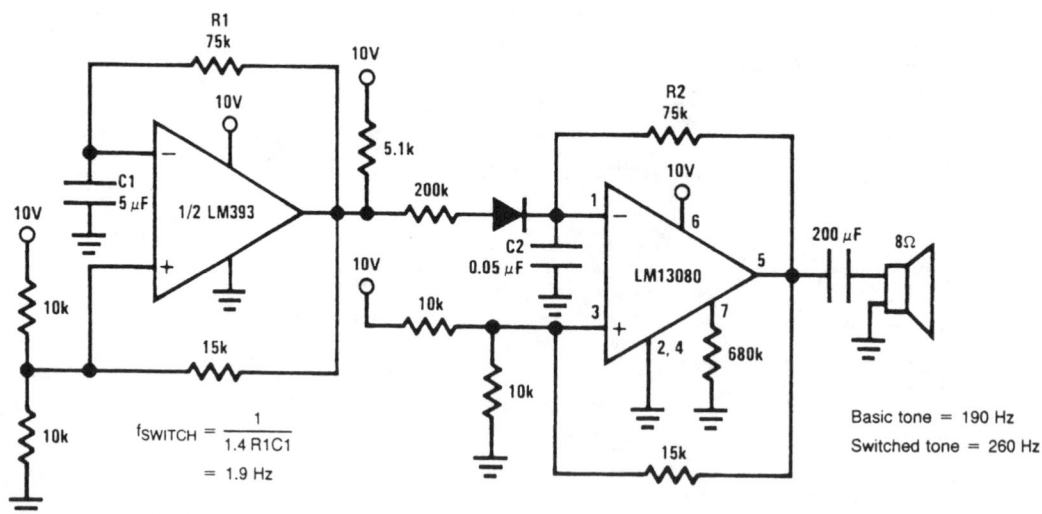

Two-Tone Siren. This is another way of accomplishing the same general performance of the previous siren circuit, while still maintaining an unnecessary degree of assembly and cost difficulties to arrive at a simple output performance.—"Linear Databook I." Courtesy of National Semiconductor Corporation, Santa Clara, California.

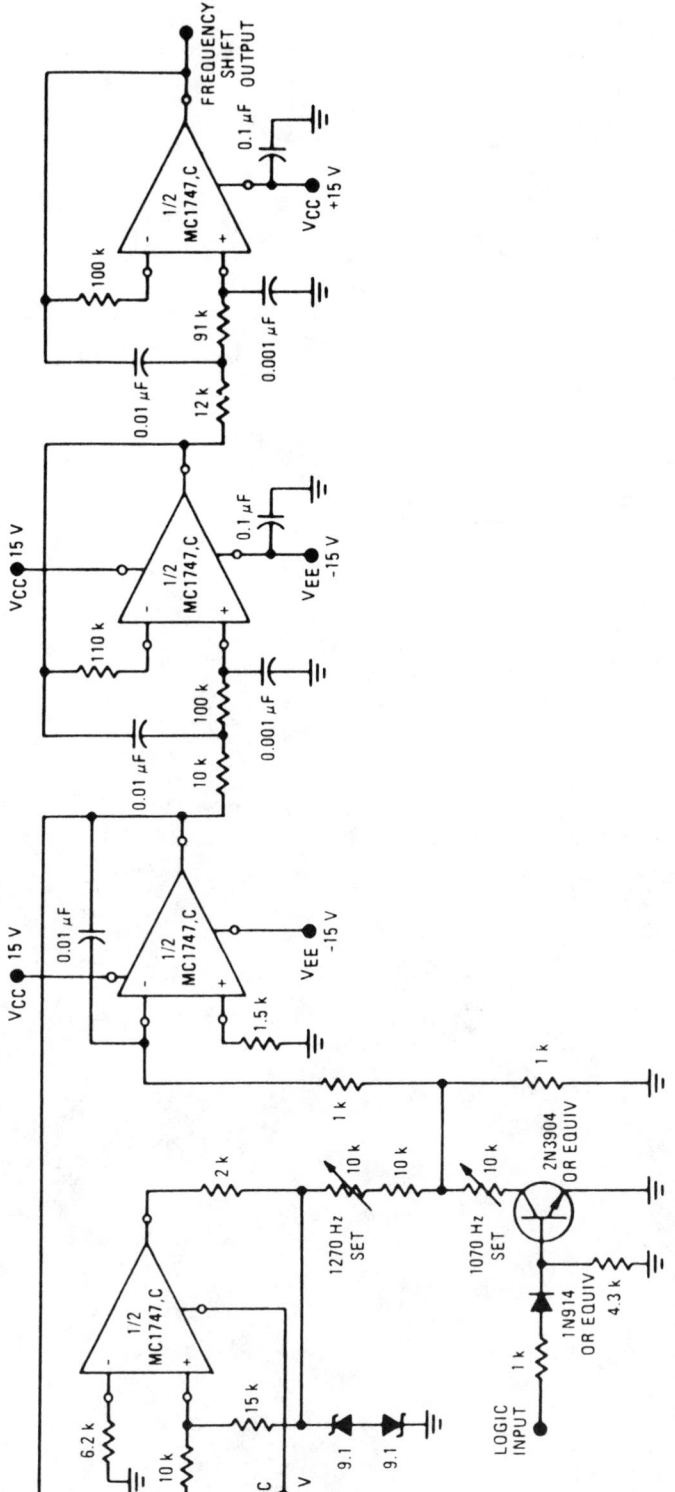

Terminals not shown are not connected.

FSK Tone Generator Test Circuit.   This circuit utilizes two Motorola MC1747C dual op amps operating from a 15-V DC bipolar supply. Two variable resistors (10 kΩ) are used to set the mark and space tones. The output is very pure and should be more than adequate for most radioteletype applications.—"Motorola Linear and Interface Integrated Circuits." Courtesy of Motorola Semiconductor Products, Motorola, Inc.

Chapter 4

Waveform Generators/Converters

The circuits found in this chapter demonstrate a very common and highly applicable use of the modern operational amplifier. Due to their inherent high-stability characteristics (when proper construction techniques are rigidly adhered to), op amps are in their element when used to construct generators that output simple and/or complex waveforms that, generally, span the audio frequency range. In these pages you will find function generators, sweep generators, triangle wave generators, and the like.

This section also includes signal/waveform converters which can be incorporated into existing laboratory equipment in order to enhance its usefulness to more critical measurement applications.

Many of these circuits are obvious expansions of and combinations of the circuits presented in the previous chapters. This can be taken a step further by combining other circuits with the same framework for extended range test generators which output stable signals of a diverse nature over an even wider range of frequencies.

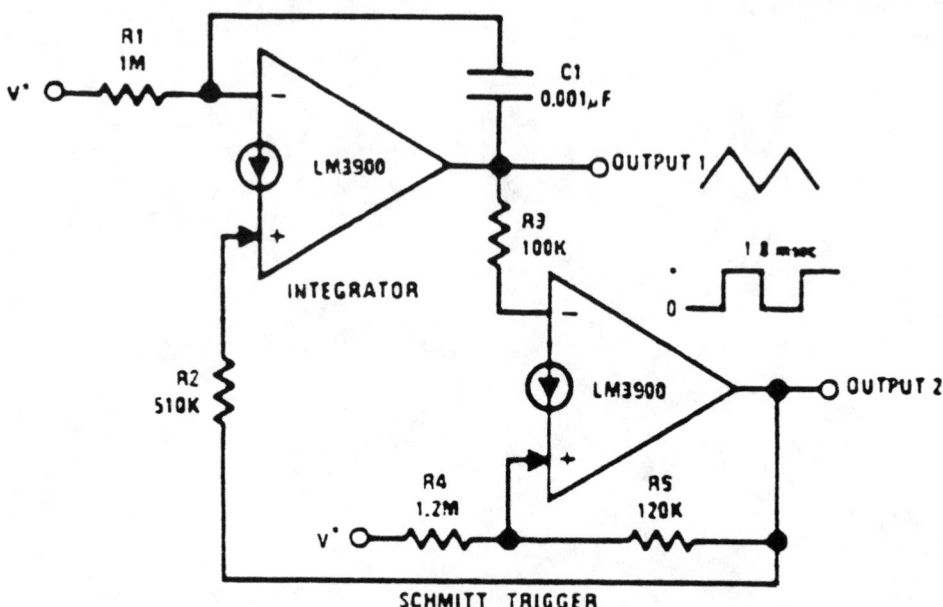

Triangle Waveform Generator. This single polarity circuit design incorporates one-half of the LM3900 quad operational amplifier to produce a triangular waveform at one output and a square wave at another. The first amplifier performs integration by operating with the current through R1. This produces a negative output voltage slope. After this occurs, the same amplifier then outputs a positive-going voltage when the Schmitt Trigger circuit (the second amplifier) goes high. The square waveform is output by the Schmitt Trigger.—"Linear Applications Databook." Courtesy of National Semiconductor Corporation, Santa Clara, California.

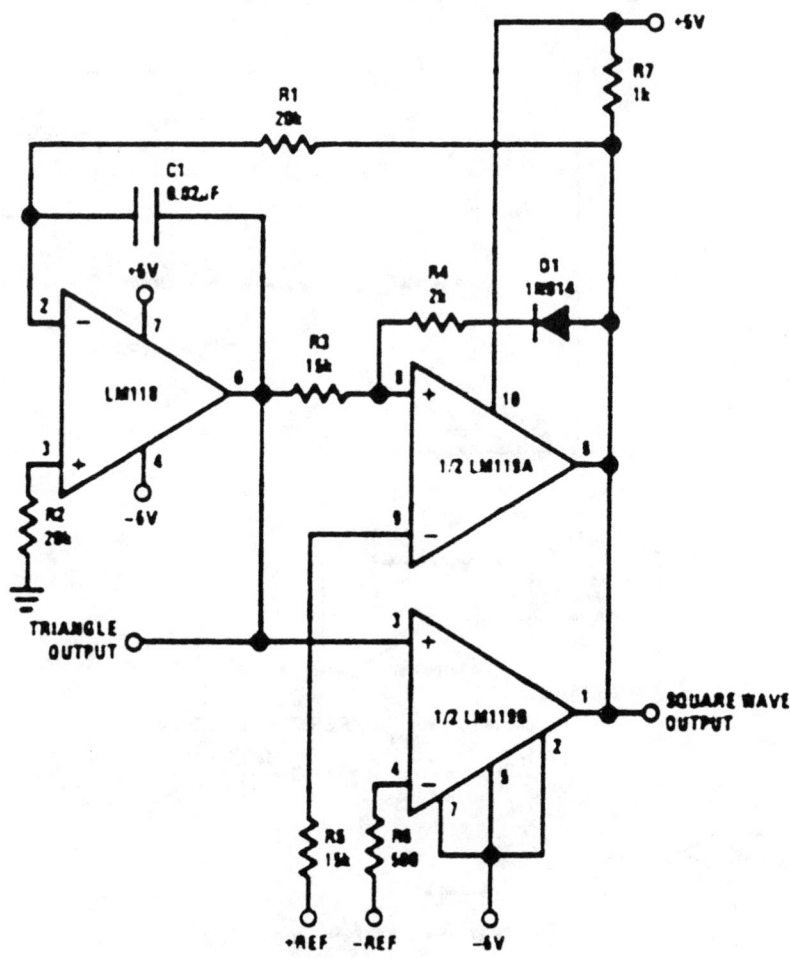

Tri-Wave Generator.   This triangle wave generator uses the LM118 op amp in conjunction with the LM119A dual comparator. The op amp is configured as an integrator and the comparators set the positive and negative peaks. Current can be injected into the inverting input of the LM118 to change ramp time.— "Linear Applications Databook." Courtesy of National Semiconductor Corporation, Santa Clara, California.

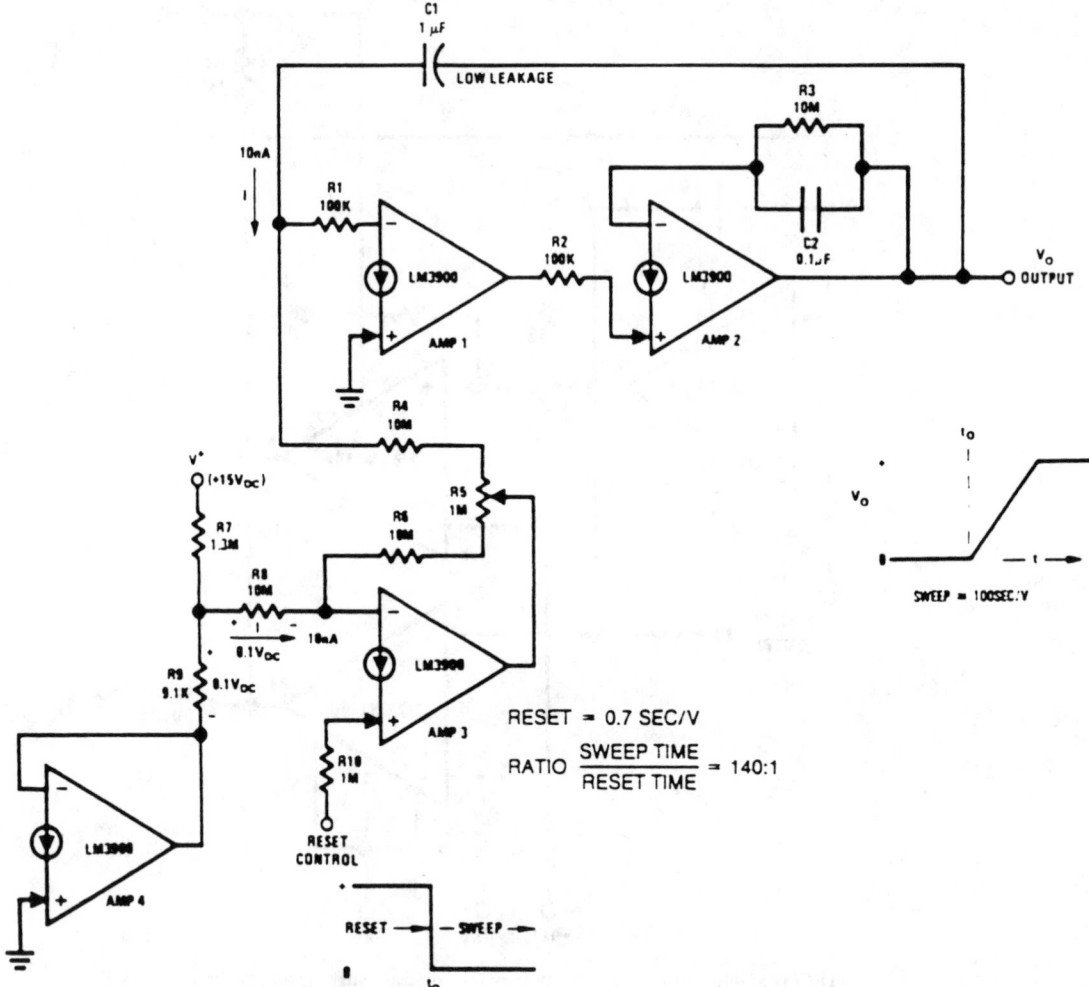

Very Slow Sawtooth Waveform Generator. Long time delay intervals can be effectively produced by utilizing a very slow sawtooth waveform as a driver. The four op amps contained in the single LM3900 package lend themselves nicely to such a circuit. Other basic sweep rates may be obtained from this circuit by scaling R8 and C1. R5 is used to minimize drift.—"Linear Applications Databook." Courtesy of National Semiconductor Corporation, Santa Clara, California.

Staircase Generator. This circuit uses three CMOS op amps. The first of the ▶ RCA CA3130 ICs is used as a multivibrator, while the second is configured as a hysteresis switch. The CA3160 is used as a linear staircase generator.—"RCA Solid State Databook—Integrated Circuits for Linear Applications." Courtesy of Harris Semiconductor.

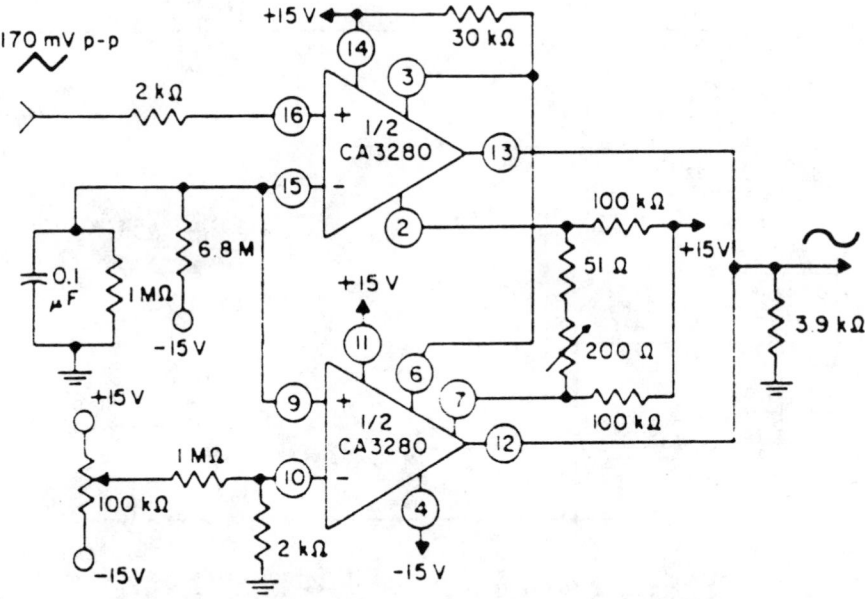

Triangle to Sine Wave Converter. This simple circuit uses the RCA CA3280 dual op amp to produce a quality (THD = 0.37%) sine wave output with a triangular wave input. Two 100-kΩ resistors are connected between the differential amplifier emitters and V+ to reduce current flow through the amplifier. This allows the amplifier to cut off fully during peak input signal excursions.—"RCA Solid State Databook—Integrated Circuits for Linear Applications." Courtesy of Harris Semiconductor.

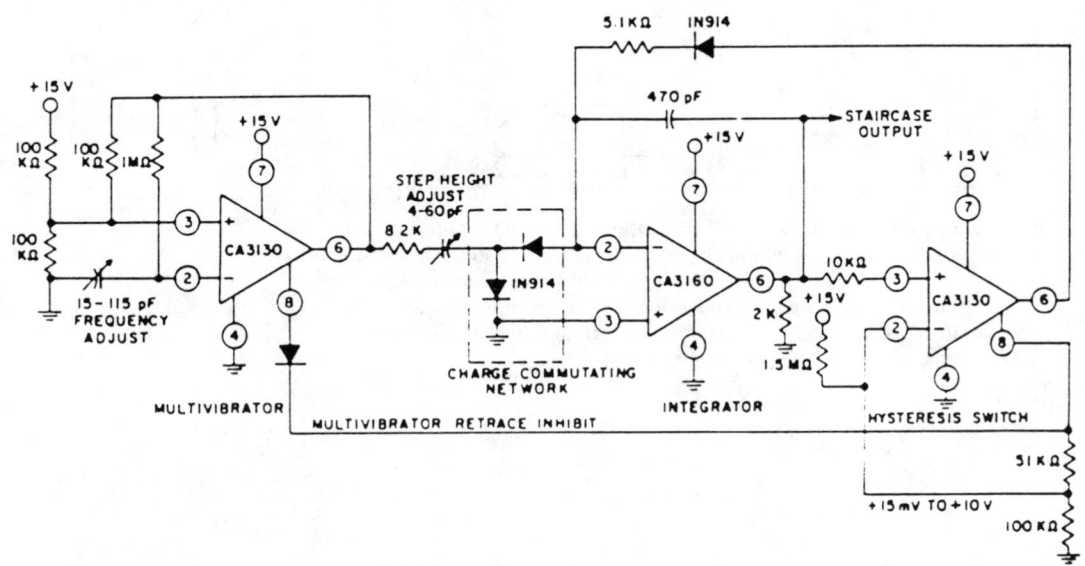

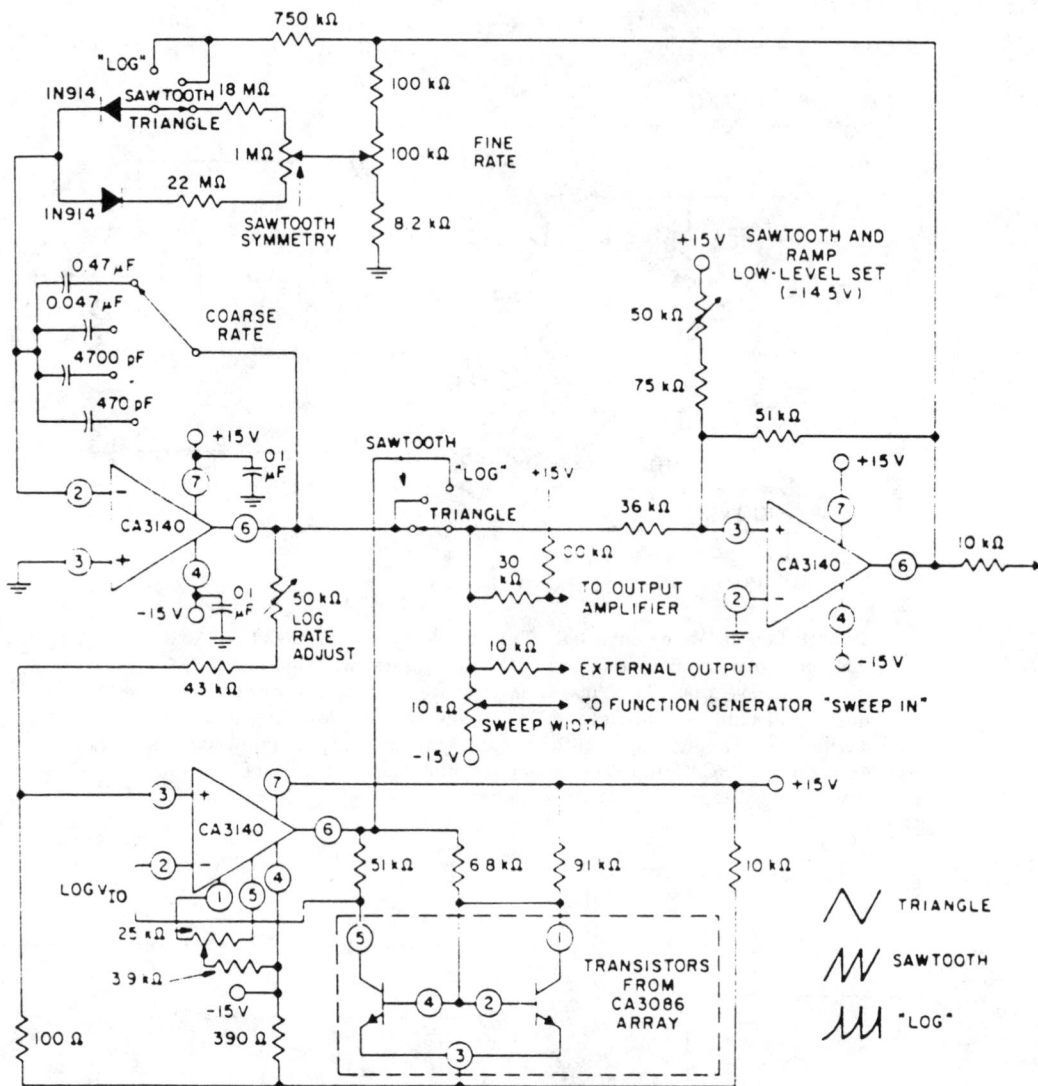

Sweep Generator. Three RCA MOSFET input/bipolar output op amps are used to form this sweep generator. One op amp is used as an integrator, another as a hysteresis switch, and the third as a logarithmic shaping network for the log function. Rates and slopes as well as logarithmic sweeps are generated by this circuit.—"RCA Solid State Databook—Integrated Circuits for Linear Applications." Courtesy of Harris Semiconductor.

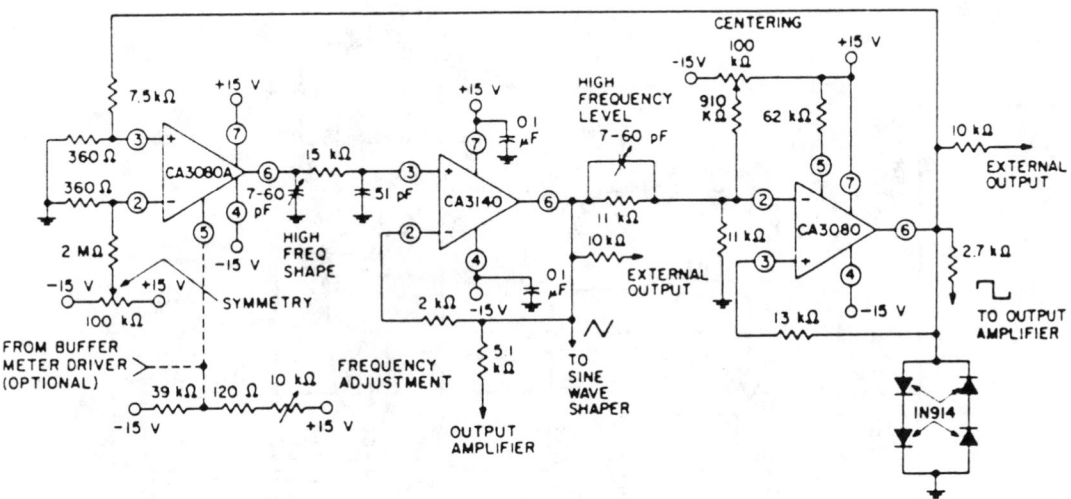

Super Sweep Function Generator. This generator has a typically wide tuning range of 1,000,000/1 and is adjusted via a single potentiometer. The CA3140 op amp functions as a noninverting read-out amplifier of the triangular signal developed across the integrating capacitor network connected to the output of the CA3080 current source. The second CA3080 receives the buffered triangular output signals, functioning as a high-speed hysteresis switch. The output from this switch is returned to the input of the original CA3080 current source, completing the positive feedback loop. Four level-limiting diodes in conjunction with the second CA3080 and a resistor–divider network determine the triangular output level. High-frequency ramp linearity is adjusted by a 7–60 pF capacitor.—"RCA Solid State Databook—Integrated Circuits for Linear Applications." Courtesy of Harris Semiconductor.

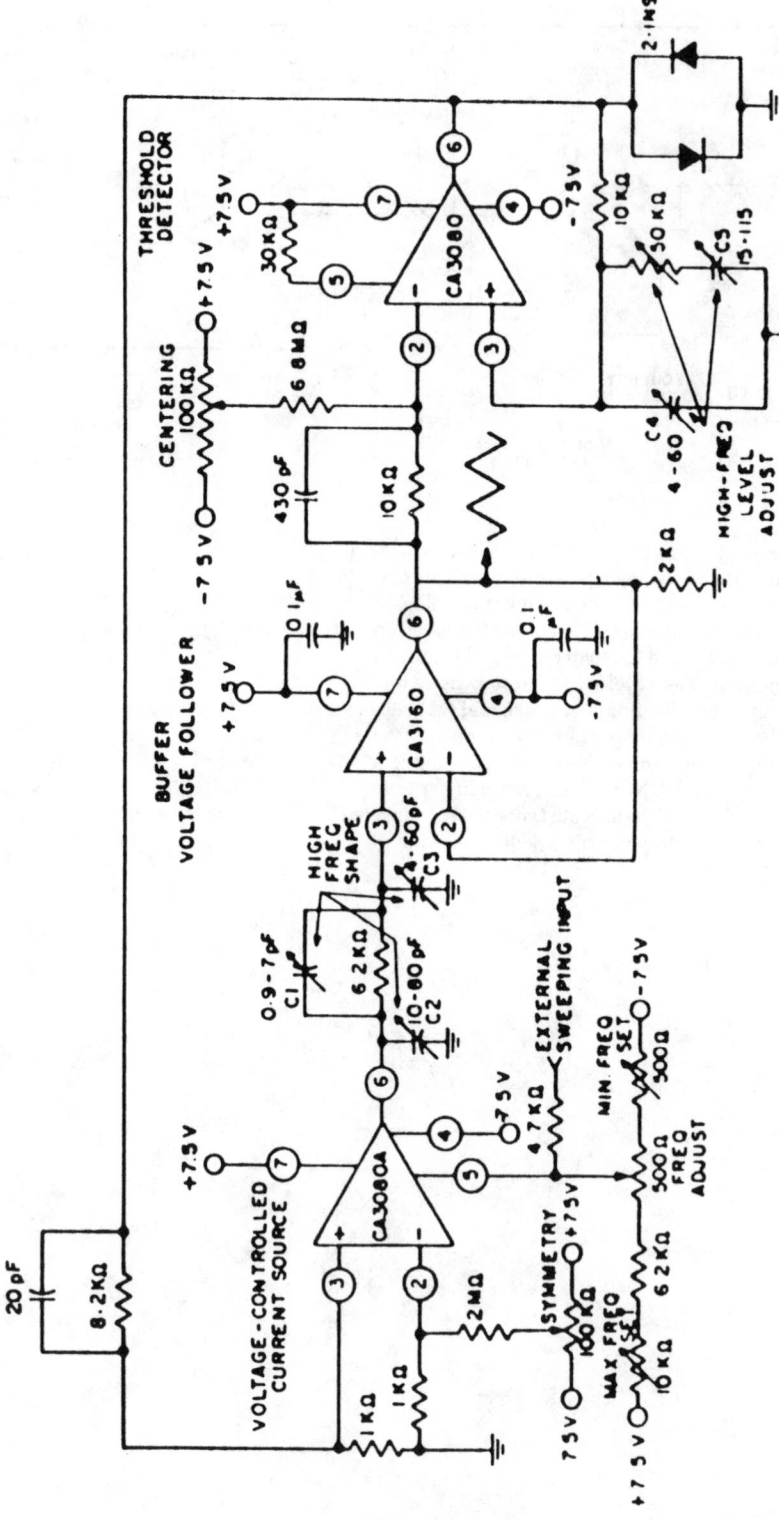

1,000,000/1 Function Generator. This circuit offers one-control adjustment of frequency over a very wide range using three op amps. Once the frequency range limits have been properly set, one control allows the frequency to be varied from a low of 1 Hz to a high of 1 MHz.—"RCA Solid State Databook—Integrated Circuits for Linear Applications." Courtesy of Harris Semiconductor.

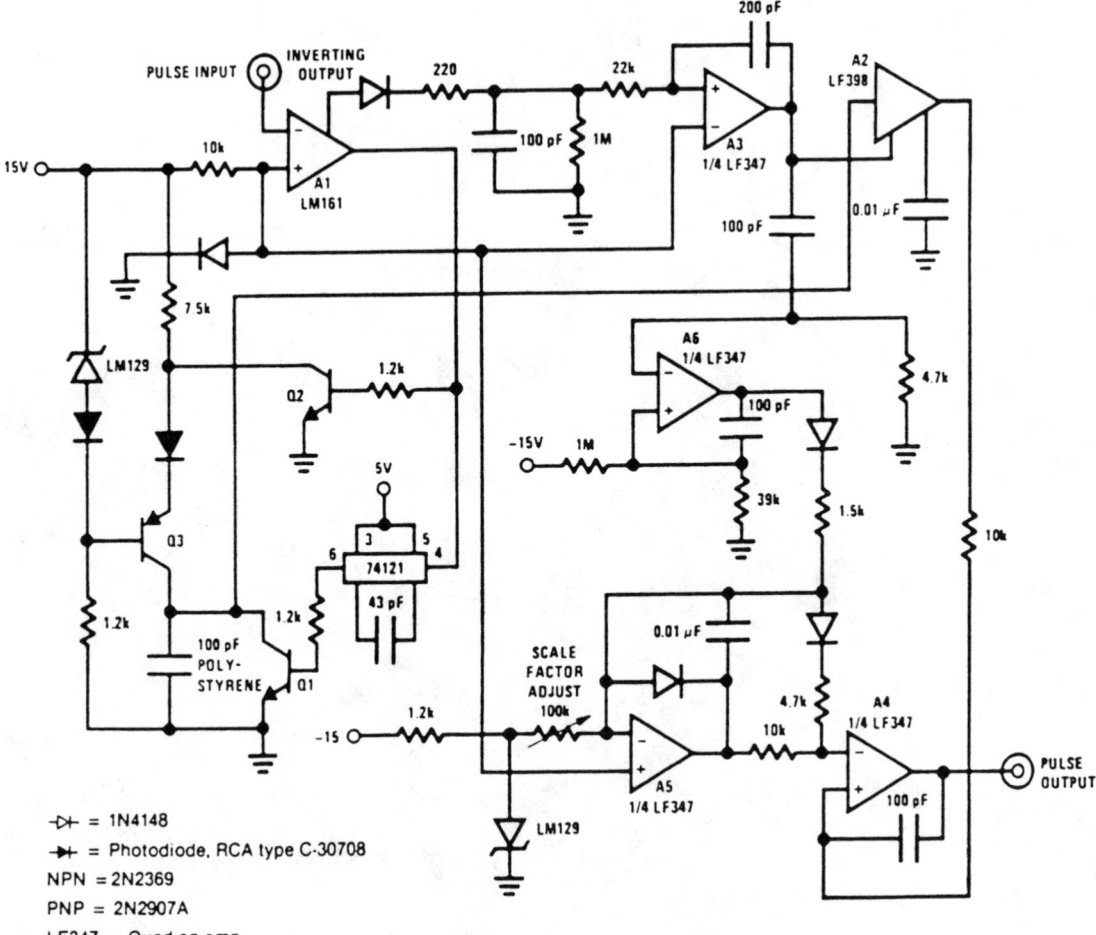

Proportional Pulse Stretcher. This complex circuit allows high-accuracy measurements of short-width pulse durations. A short input pulse triggers the 74121 one-shot and discharges the 100 pF capacitor through Q1. At the same time Q3, the recharging current source, is turned on. As long as the input pulse is present, the capacitor charges. When the pulse ends, the capacitor's voltage is proportional to the width of the pulse. A2, the sample and hold amplifier, samples the voltage, and the resultant DC level controls the "on" duration of the pulse width modulator configuration of the LF347 op amp. The time amplification factor for this circuit is approximately 2000 with a 1 μsec full-scale width giving a 1.4 msec output pulse.—"Linear Applications Databook." Courtesy of National Semiconductor Corporation, Santa Clara, California.

Chapter 5

Test Instruments

Operational amplifiers now form the heart of many different types of test or "bench" instruments. The excellent stability of the modern IC op amp lends itself well to such instruments, where accuracy of measurements is highly critical.

This chapter highlights many of these instruments, ranging from highly accurate ohmmeters that can be constructed with a bare minimum of different components, extremely sensitive voltmeters and ammeters, and other instruments that until a few years ago would have consumed a great deal of construction space. Of course, highly stable instruments of this nature have been available for quite a long time, but most of those contained in this section can be constructed from a small number of components and for a fraction of the cost of the commercial offerings.

More importantly, these circuits can be used as basic building blocks for more sensitive measuring equipment or as the core of a multi-testing device which might meet the specific needs of a critical measuring application for which there is no commercial instrument.

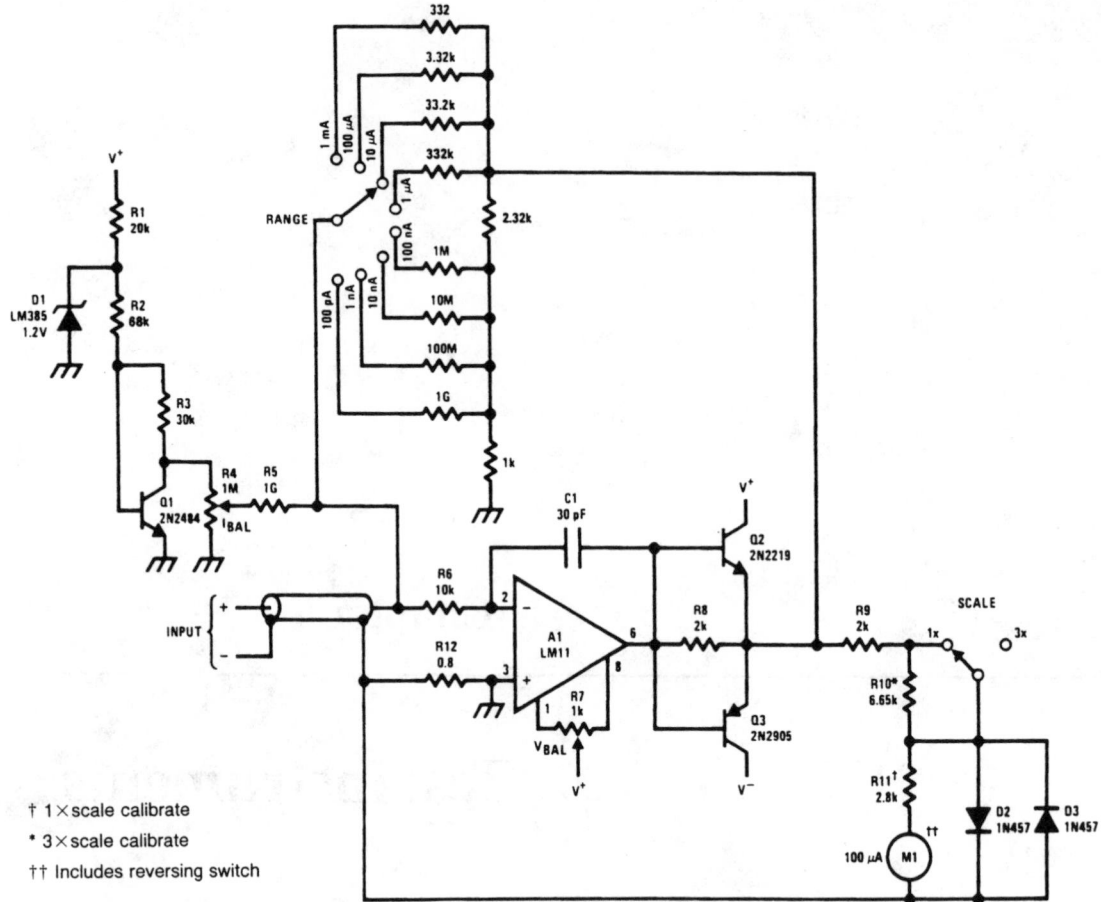

Wide-Range Low-Level Ammeter.   This ammeter circuit is designed to produce excellent accuracy over a wide range of currents from 100 pA to 3 mA. It uses a summing amplifier connection to minimize the voltage drop across input terminals. A floating power supply is incorporated, so the power ground and the signal ground are separated by resistor R12. Bias current compensation is used to increase the meter sensitivity. This means that there are two meter zeroing adjustments. Potentiometer R4 allows for current balancing which is used most effectively while in the most sensitive ranges. Voltage balancing is provided by R7 and is used below 100 μA and with the inputs shorted. The LM11 op amp has internal limiting via back-to-back diodes at terminal 2. Current-limiting resistor R6 is added at this terminal to protect against overloads. This resistor may be omitted if desired, but it does not effect accuracy and is highly recommended in metering applications where the current levels may approach or exceed 10 mA.—"Linear Databook I." Courtesy of National Semiconductor Corporation, Santa Clara, California.

Although a few of the components in some of these circuits must be of precision quality, most should be readily available at any electronic supply outlet. The builder's accuracy requirements will dictate the tolerances of many of these, and less stringent tolerances may be chosen for simple test instruments on which lesser demands are placed.

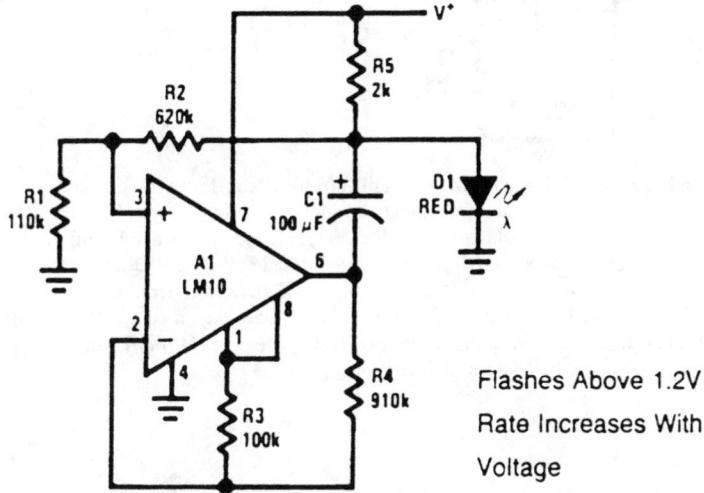

Flashes Above 1.2V

Rate Increases With

Voltage

Single-Cell Voltage Monitor.   This is an interesting circuit, although its applications may be a bit limited. The LED is triggered at a rate that is voltage-dependent. Below about 1.2 V DC, the LED is not triggered. Activation occurs at potentials above 1.2 V DC and the flash rate increases as the voltage level rises. This circuit is reminiscent of the days when "crude" indicators were the norm, taking the place of the then expensive d'Arsonval meter. However, it might offer some potential in a single-cell battery charger circuit, in that most of the rechargeable cells produce a 1.2 V DC output at full charge. When the LED begins flashing, the battery is charged to its full potential. The circuit that follows is a more useful rendition of this one.—"Linear Databook I." Courtesy of National Semiconductor Corporation, Santa Clara, California.

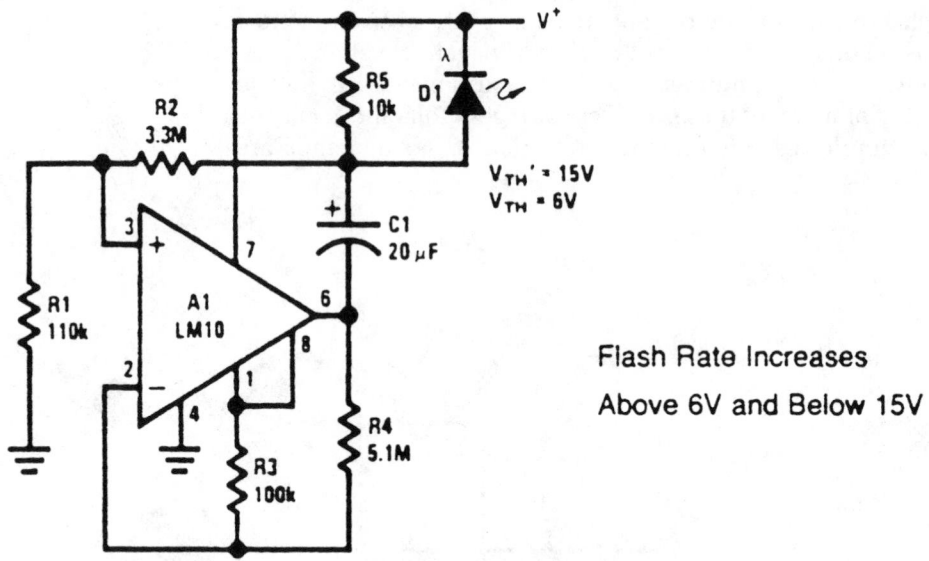

Flash Rate Increases

Above 6V and Below 15V

Double-Ended Voltage Monitor. This circuit is similar to the previous one; however, this one is used to indicate a range of voltage values at the input to the op amp. Here, the flash rate of the LED increases above 6 V DC and below 15 V DC. The most rapid triggering occurs at just shy of 15 V DC. Beyond this point the flash rate drops off. Both circuits utilize current from the single polarity voltage source they monitor and are, obviously, designed for battery level applications.—"Linear Databook I." Courtesy of National Semiconductor Corporation, Santa Clara, California.

Picoammeter.   This circuit incorporates an inexpensive 500–0–500 micro- ▶ ammeter to provide a picoammeter circuit with a ± 3 pA full-scale reading. The CA3140 op amp serves as a times 100 gain stage to boost the low current levels being measured to a point where they may be indicated by the micro-ammeter.—"RCA Solid State Databook—Integrated Circuits for Linear Appli-cations." Courtesy of Harris Semiconductor.

Low-Current Ammeter.   This is an
excellent circuit that offers economy and
good accuracy. It is preferable to use a
precision trimmer potentiometer for meter
scale adjustment. The dual voltage power
supply can take the form of two 9-V
batteries for portable operation. For use at
the bench a dual 12–15-V supply with
excellent dynamic regulation
characteristics is necessary.—"Linear
Databook I." Courtesy of National
Semiconductor Corporation, Santa Clara,
California.

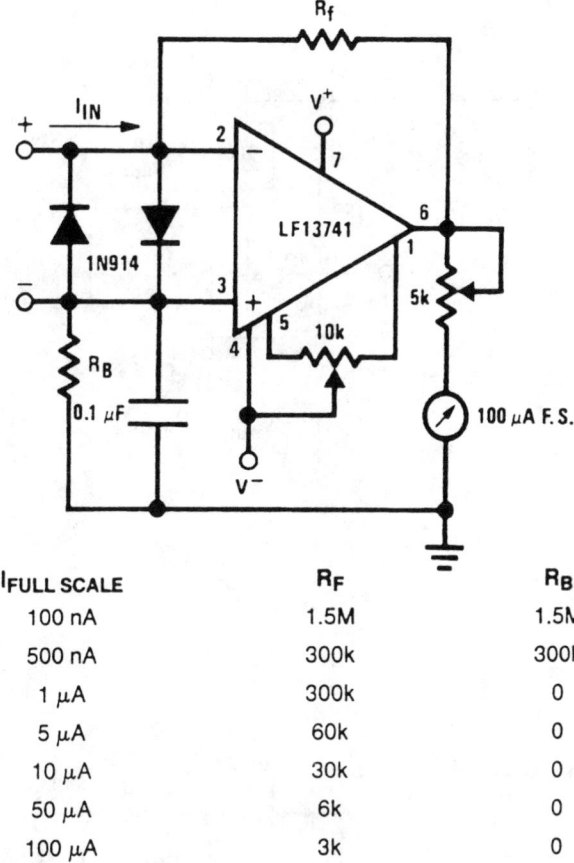

$I_{FULL SCALE}$	R_F	R_B
100 nA	1.5M	1.5M
500 nA	300k	300k
1 μA	300k	0
5 μA	60k	0
10 μA	30k	0
50 μA	6k	0
100 μA	3k	0

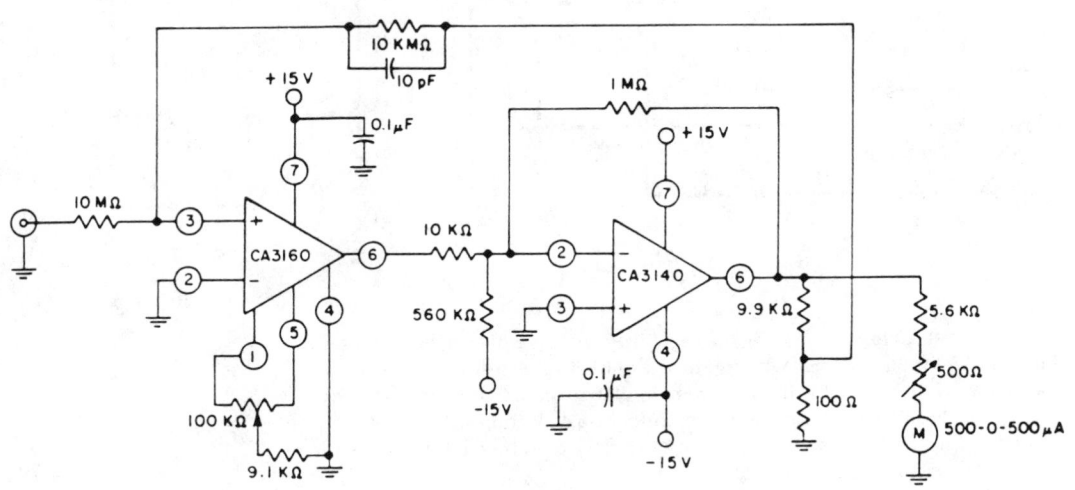

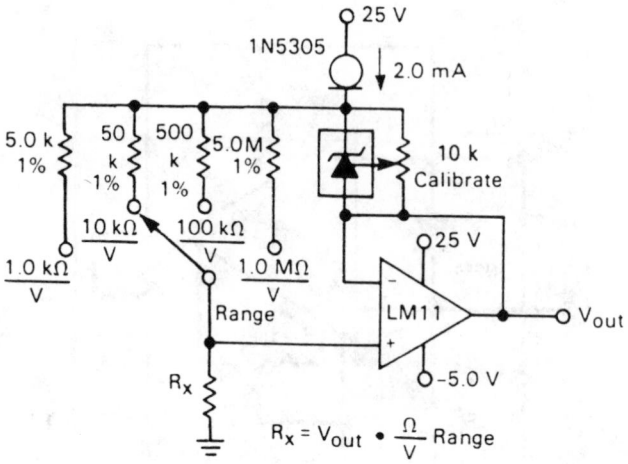

Linear Ohmmeter. Utilizing the Motorola TL431A series three-terminal programmable shunt regulator diodes, this ohmmeter offers scales ranging from 1000 Ω to 5 MΩ. The 10 kΩ potentiometer should be a linear taper, precision type. The range switch should be a high-quality ceramic type.—"Motorola Linear/Switchmode Voltage Regulator Handbook." Courtesy of Motorola Semiconductor Products, Motorola, Inc.

$$R_x = V_{out} \cdot \frac{\Omega}{V} \, \text{Range}$$

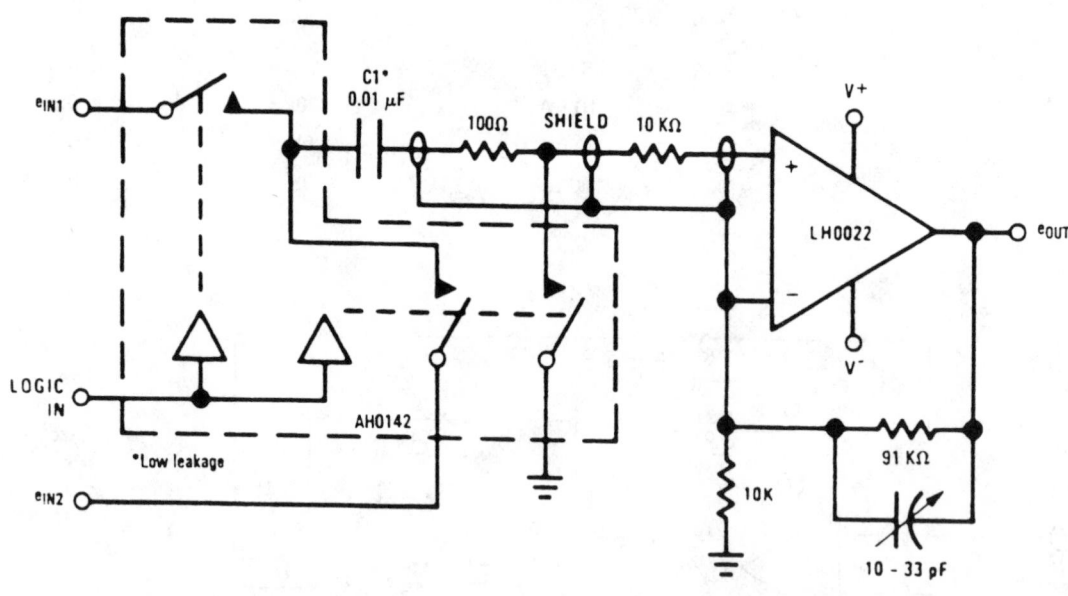

$$e_{OUT} = 10 \times (e_{IN1} - e_{IN2})$$

Low-Current Ammeter. Having the capability of measuring current values in the 100-pA to 100-μA range without the use of high value resistors, this circuit uses the LM216A ultra low input current bipolar op amp. Calibration requires only a single adjustment. Adjust R4 for a full-scale deflection with a 1-μA input.—"Linear Applications Databook." Courtesy of National Semiconductor Corporation, Santa Clara, California.

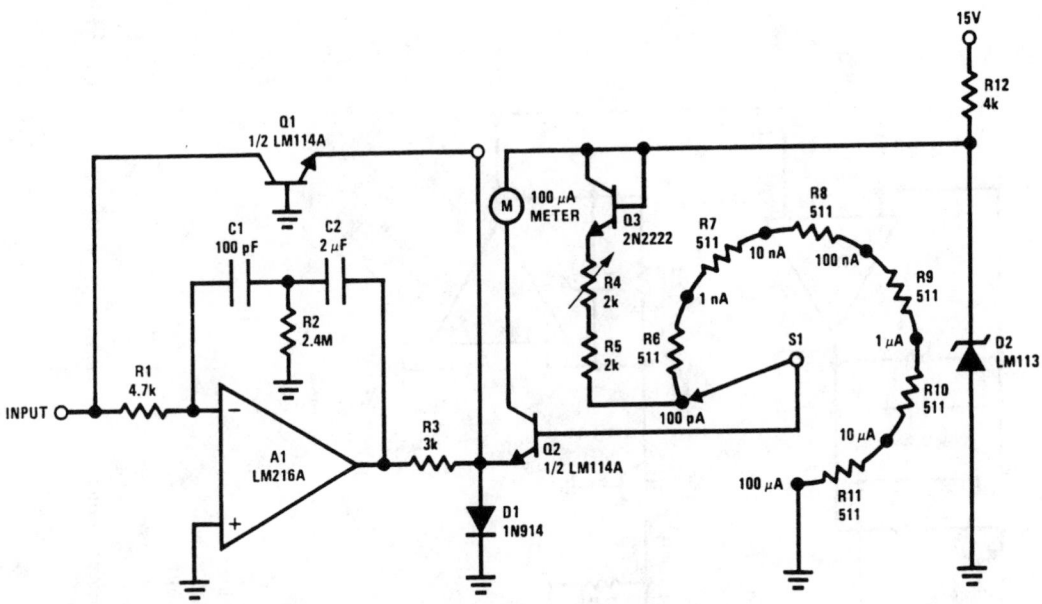

Sensitive Low-Cost "VTVM."   This is a good voltmeter circuit that can be constructed from "junk box" parts with the exception of the 1% resistors which will probably have to be purchased. 1N914 diodes may be used between pins 2 and 3, or any small signal types may be substituted in this limiter configuration. Use a high-quality ceramic rotary switch to avoid measurement inaccuracies as the components age. The LH0042H op amp is an FET input type that offers excellent sensitivity and may be operated at supply voltages up to 22 V DC.—"Linear Databook I." Courtesy of National Semiconductor Corporation, Santa Clara, California.

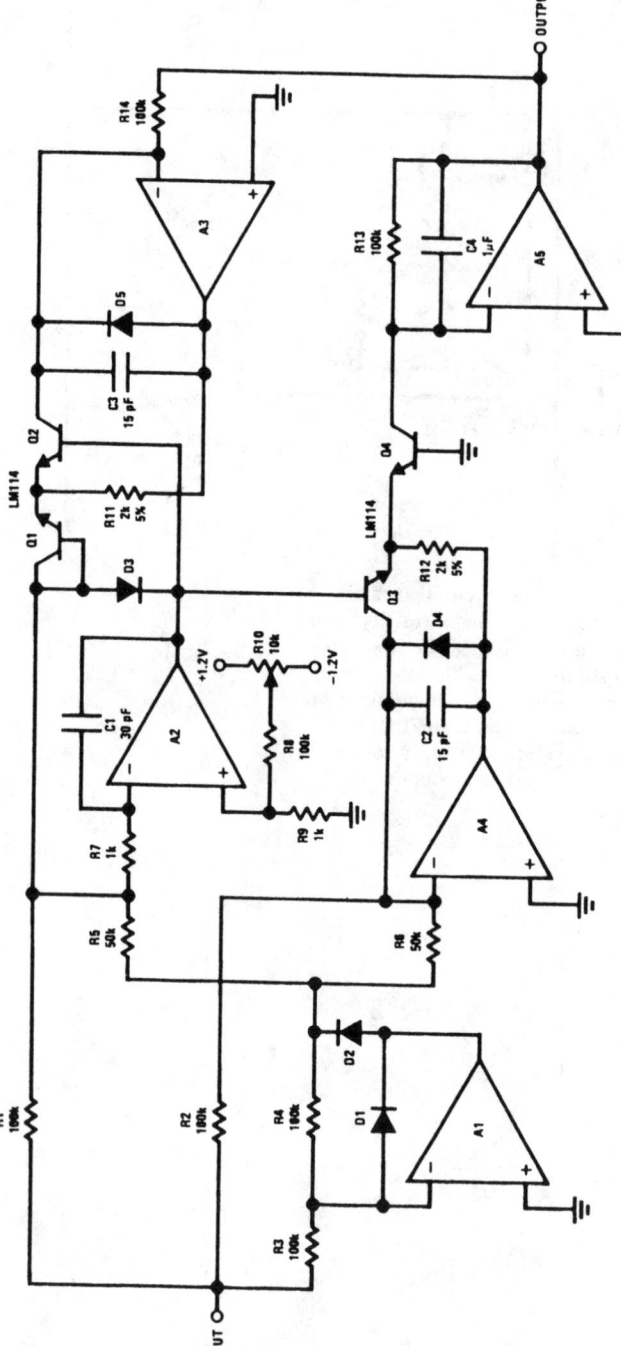

Note 1: All operational amplifiers are LM118.

Note 2: All resistors are 1% unless otherwise specified.

Note 3: All diodes are 1N914.

Note 4: Supply voltage ±15V.

True RMS Detector.   This is an op amp precision rectifier circuit that makes it possible to measure millivolt AC signals with a DC meter while maintaining an accuracy of better than 1%. The basic configuration is that of a precision absolute value circuit coupled to a one-quadrant multiplier/ divider. For best results transistors Q1 through Q4 should be matched, have high-beta, and be operated at the same temperature. Dual transistor packages might be considered for this application. —"Linear Applications Databook." Courtesy of National Semiconductor Corporation, Santa Clara, California.

Wide-Range AC Voltmeter.
This is a circuit that has
personally served the author
for many years. It can be
constructed in a very short
period of time and very
economically, since it uses
components that are *very*
standard. Precision resistors
used in the switching circuit
will enhance accuracy, but
"junk box" components will
yield surprisingly good
results. The bridge rectifier
output to the meter can be
comprised of individual
diodes, but a single bridge
component package is
recommended for ease of
construction and to keep size
to a minimum. Make certain
all component leads are cut
to absolute minimum length
for best frequency
response.—"Linear Databook
I." Courtesy of National
Semiconductor Corporation,
Santa Clara, California.

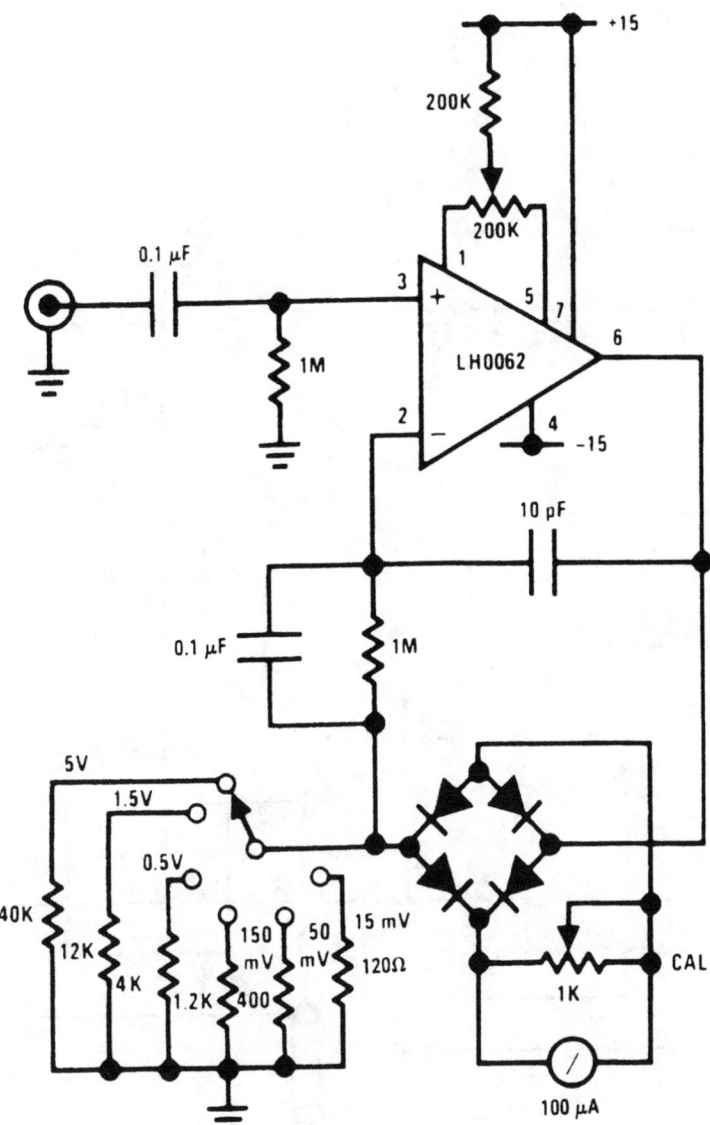

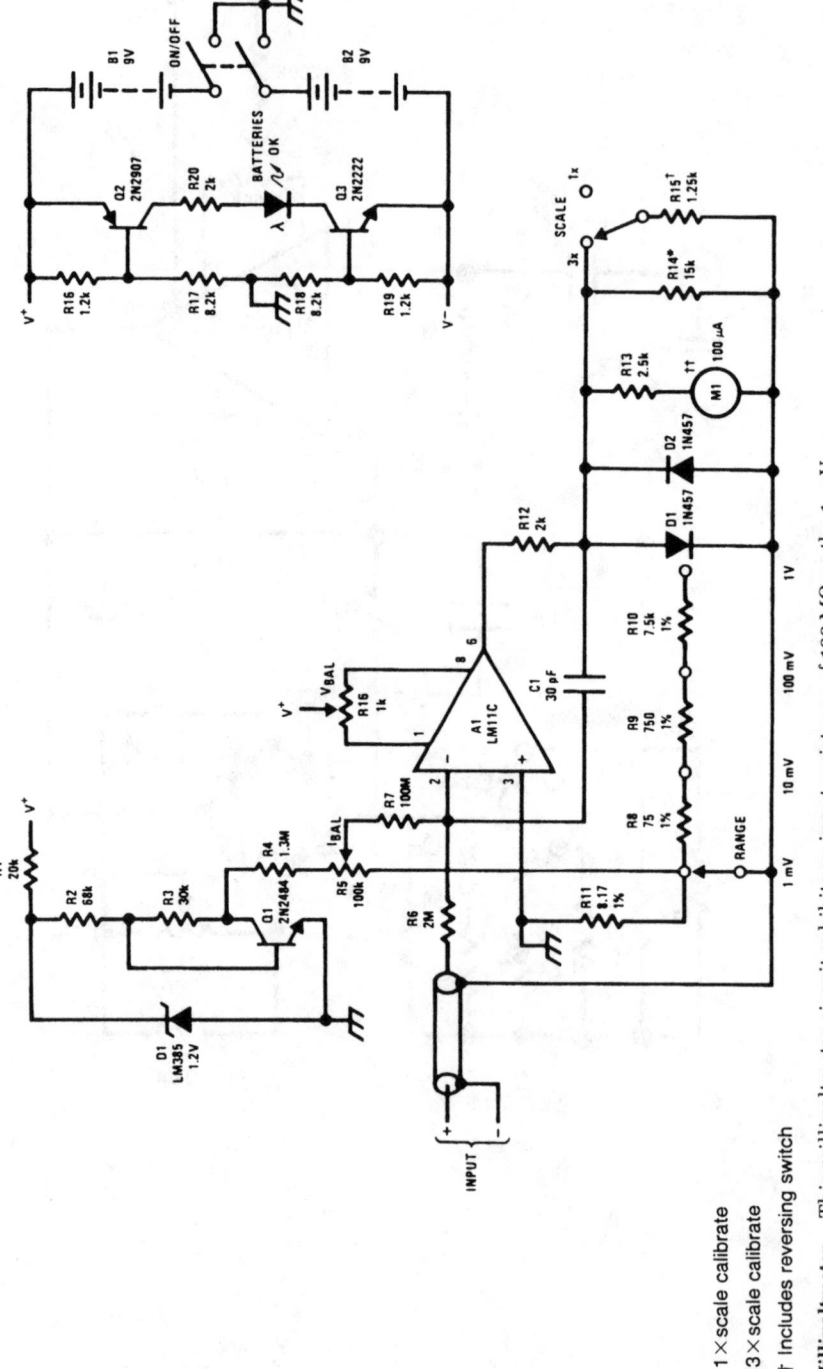

* 1×scale calibrate

† 3×scale calibrate

†† Includes reversing switch

Millivoltmeter. This millivoltmeter circuit exhibits an input resistance of 100 MΩ on the 1-mV scale and rising to 300 GΩ on the 3-V scale. Notice that the power and signal grounds are separated in order to facilitate bias current compensation. If this were not done, a separate op amp would be required to bootstrap the compensation to the input. The op amp is protected from overload by input resistor R6. This also ensures that an overloaded input will not result in an erroneous meter indication. Two pairs of 9-V batteries in series provide dual polarity power to this circuit, and a built-in battery condition indicator monitors the power supply during operation. Another built-in feature is found in the addition of resistors R14 and R15 which are used for calibrating the 1 × and 3 × scales, respectively.—"Linear Databook I." Courtesy of National Semiconductor Corporation, Santa Clara, California.

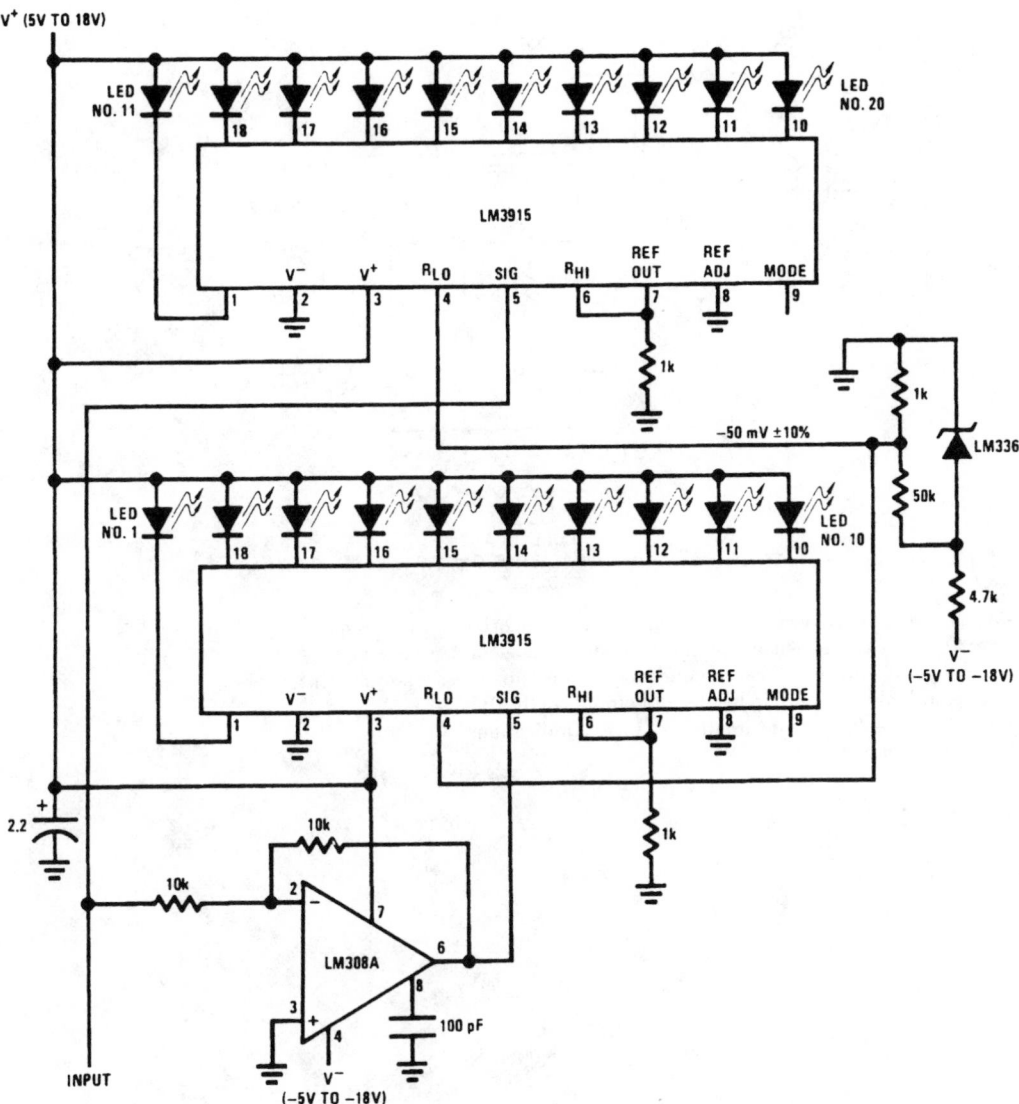

Precision Null Meter. This circuit uses the output from an LM108A op amp to drive two LM3915 Dot/Bar Display Drivers. A bipolar power supply is necessary, but values of 5 to 18 V DC allow for the possibility of incorporation in existing circuitry.—"Linear Databook III." Courtesy of National Semiconductor Corporation, Santa Clara, California.

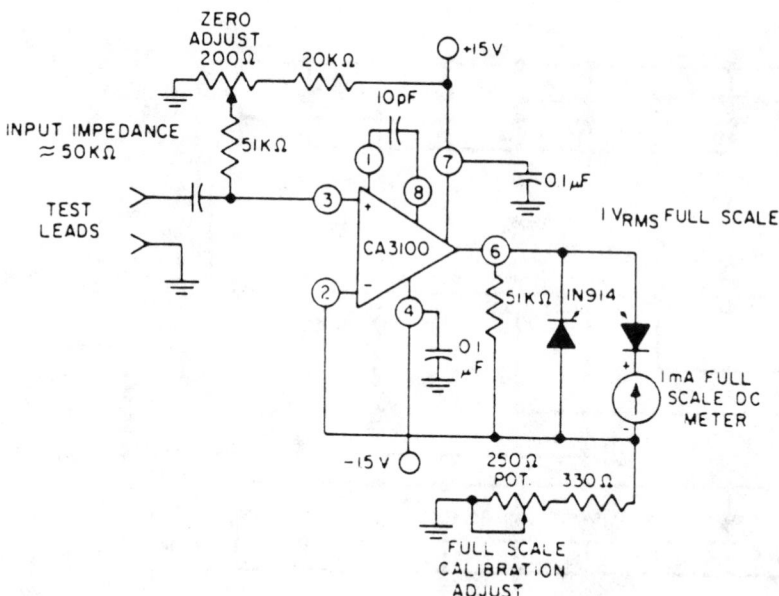

1-Megahertz Meter Driver. A wideband op amp is incorporated in this meter driver that uses an inexpensive DC milliammeter with a 1-mA scale. Make certain the 250- and 200-Ω pots are linear taper, precision types. Many surplus units are available that would be ideal for circuits of this type—"RCA Solid State Databook—Integrated Circuits for Linear Applications." Courtesy of Harris Semiconductor.

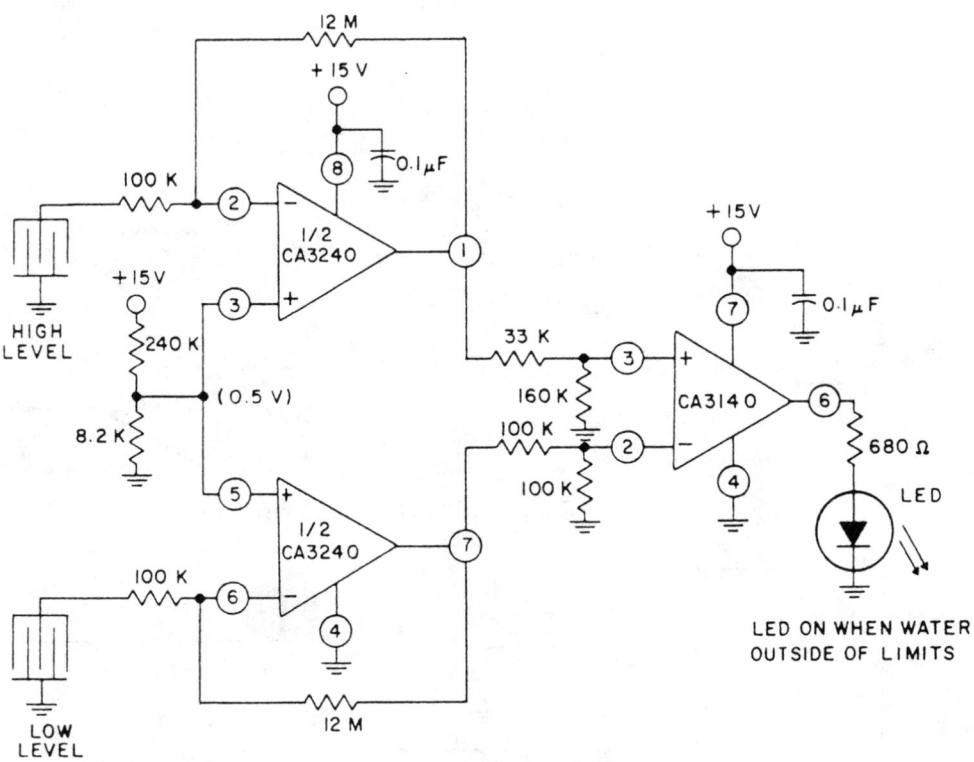

Dual Level Detector. This circuit is used to measure liquid levels in two different containers and triggers an LED when the levels are different. The RCA CA 3240 is a dual op amp that is the dual equivalent of the CA3140 also used in this circuit. This is a basic window comparator circuit that depends upon the principle that most liquids contain enough ions in solution to sustain a small amount of current flow between the printed circuit board grids that make up the level indicators.—"RCA Solid State Databook—Integrated Circuits for Linear Applications." Courtesy of Harris Semiconductor.

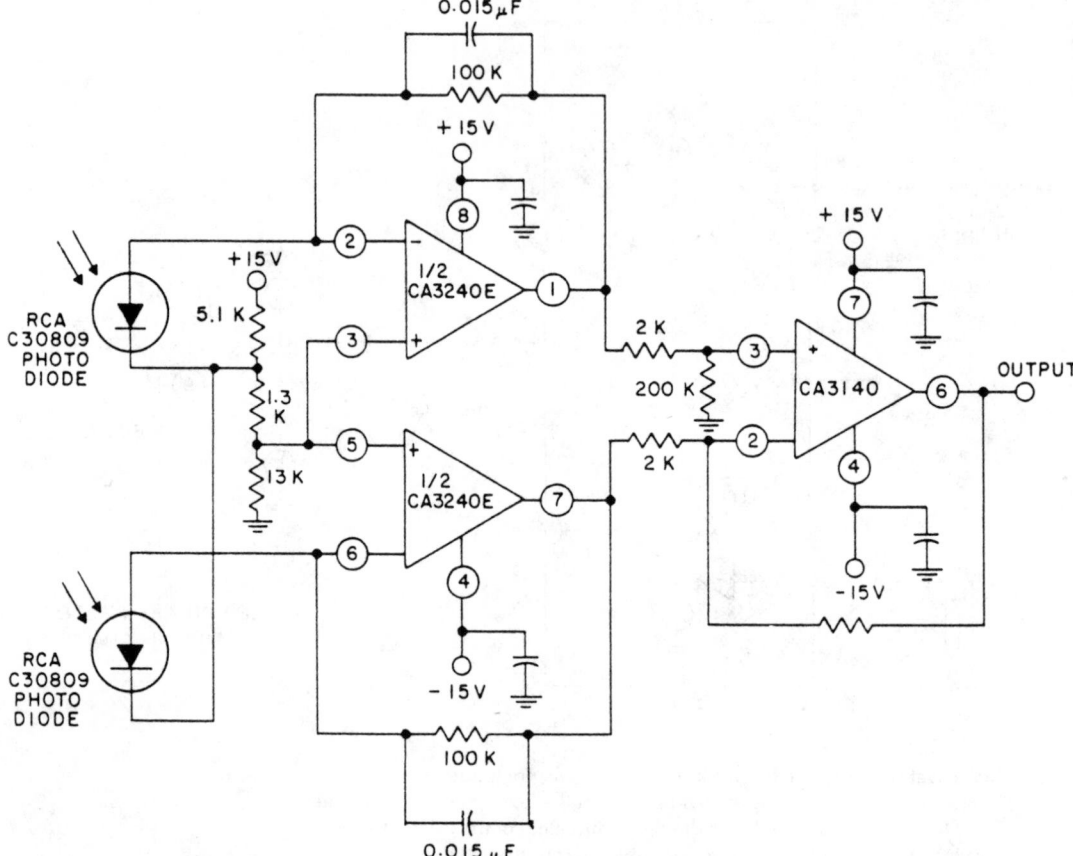

Differential Light Detector.   This circuit monitors the light levels of two
different sources and amplifies the difference voltages. This is a familiar dual
channel circuit that can accommodate many different types of input devices,
which in this case are photodiodes.—"RCA Solid State Databook—Integrated
Circuits for Linear Applications." Courtesy of Harris Semiconductor.

Peak Detectors (*two schematics*). These two circuits are used to protect (a) positive-going and (b) negative-going peaks. With large signal inputs, the bandwidth of the positive peak detector circuit offers a far wider bandwidth than does its negative-going counterpart. The latter scopes in at about 300 kHz compared with over 1 MHz for the former.— "RCA Solid State Databook— Integrated Circuits for Linear Applications." Courtesy of Harris Semiconductor.

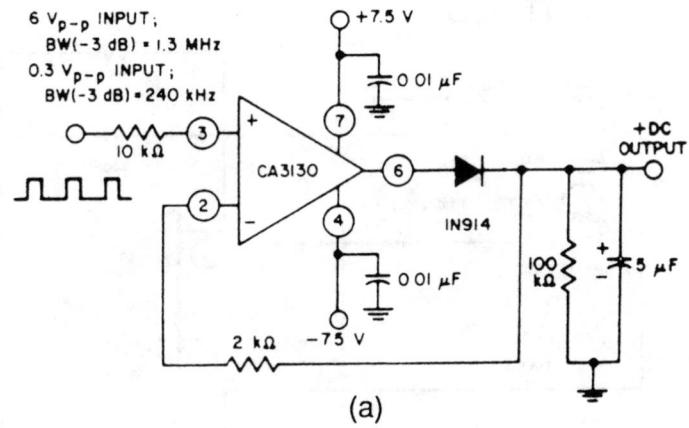

(a)

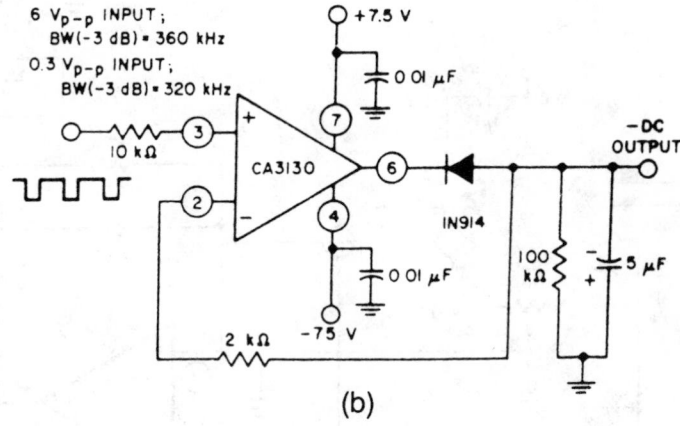

(b)

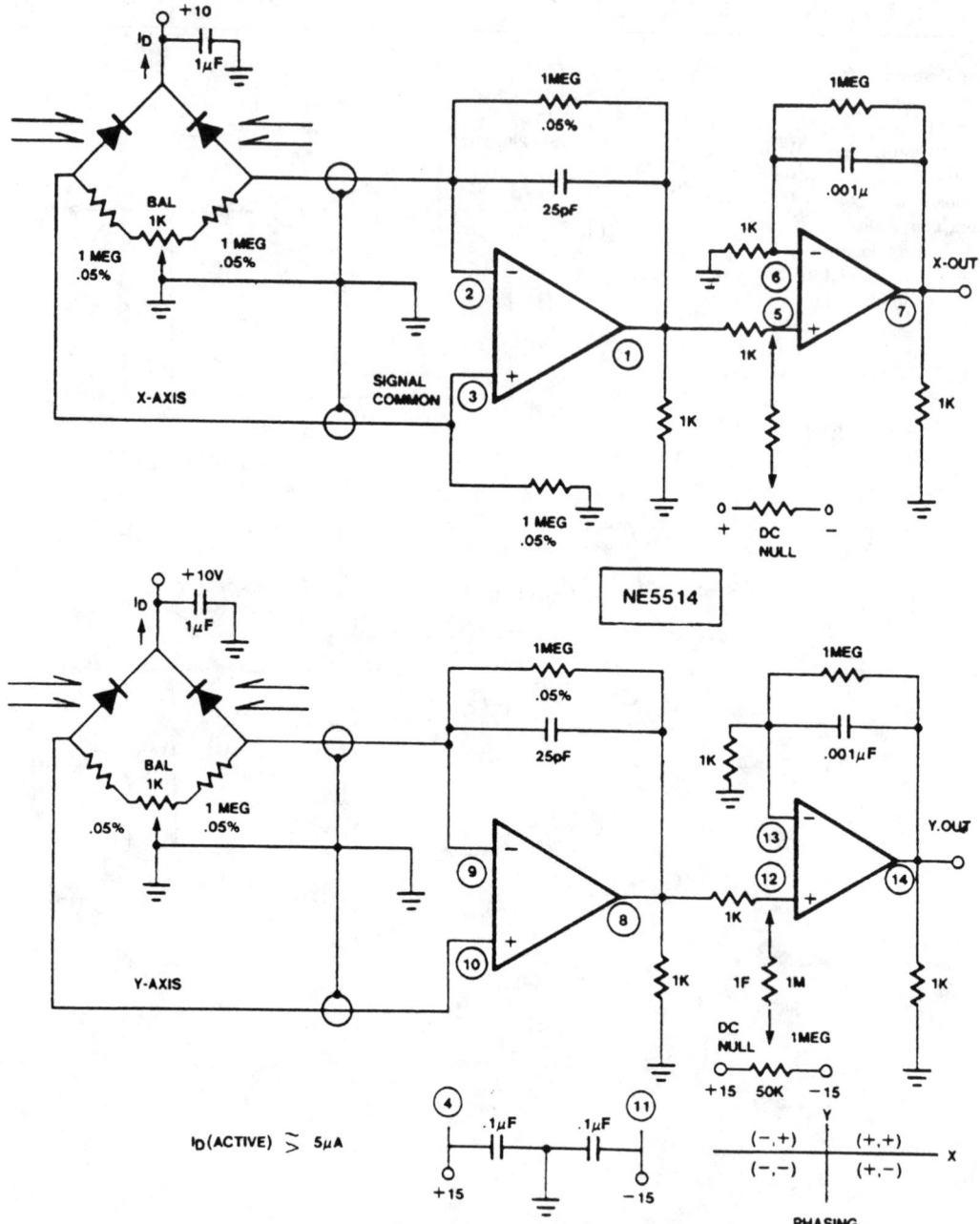

Four-Quadrant Photo Detector. This circuit senses extremely small currents from a photodiode operating in the photoconductive (reverse-biased) mode. The Signetics NE5514 quad operational amplifier has sufficiently low input bias current (about 6 nA) to allow its use under conditions that can cause inaccuracies due to the level of input bias currents. The wide input common-mode voltage range allows for a single polarity +10 V DC supply to be used to drive the signal bridge. This results in much improved signal-to-noise ratios and high sensitivity.—"Linear Data Manual Volume 2: Industrial." Courtesy of Signetics Company, a division of North American Philips Corporation.

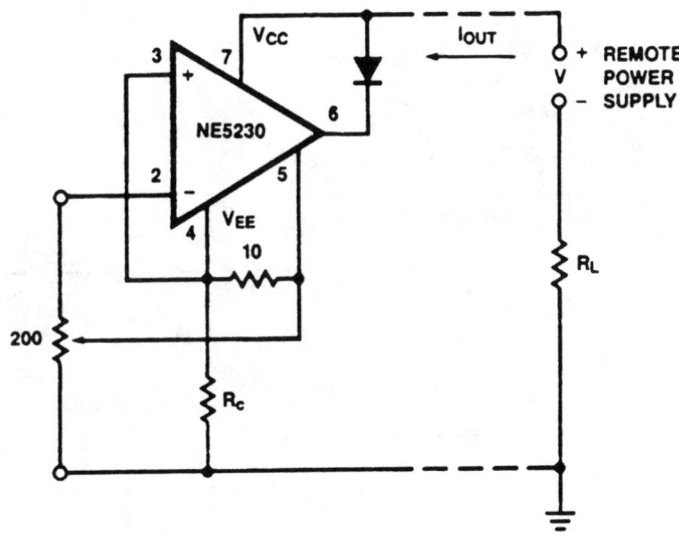

NOTES:

1. $I_{OUT} = V_{IN}/RC$

2. $R_L Max \simeq \dfrac{V_{REMOTE} - 1.8V - V_{IN}Max}{I_{OUT}}$

For RC = 1Ω

$\dfrac{I_{OUT}}{}$	$\dfrac{V_{IN}}{}$
4mA	4mV
20mA	20mV

Remote Temperature Transducer. This is a PTAT circuit meaning that the ParT itself is A Transducer. This circuit makes use of the supply current control pin, as the voltage at this pin is proportional to the absolute temperature. That is, it is produced by the amplifier bias current through an internal resistor–divider. While the author has not personally experimented with this particular circuit, it should be noted that most attempts to produce reliable, practical results with the PTAT concept have amounted to little more than an exercise in futility.—"Linear Data Manual Volume 2: Industrial." Courtesy of Signetics Company, a division of North American Philips Corporation.

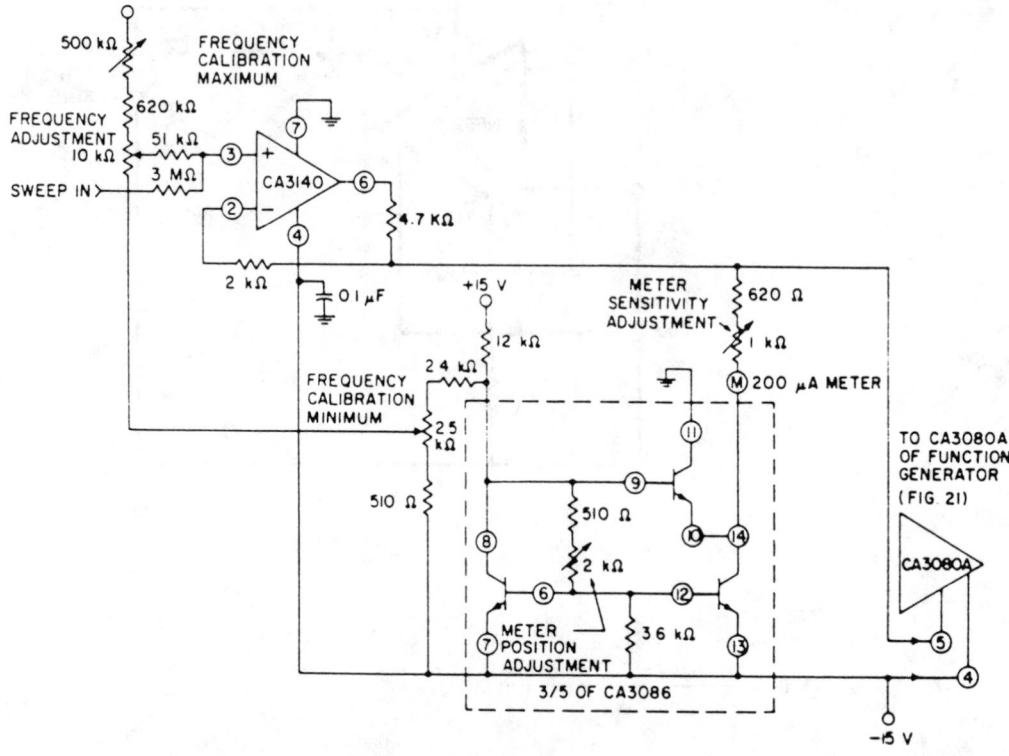

Meter Driver/Buffer Amplifier. This circuit uses an op amp connected as a voltage follower to assure a low impedance drive for smooth operation. To calibrate this circuit, set the 10-kΩ frequency adjustment potentiometer to its low end and then adjust the frequency calibration minimum control for the lowest frequency. Next, set the first pot at its upper limit, then set the maximum frequency adjust for the maximum frequency. These controls will interact, so minor readjustment will be necessary. The meter sensitivity control is used to adjust the meter scale with each decade, while the meter position control sets the pointer on the scale with minimal effect on the sensitivity. Note: An analog read-out is easily obtained from this circuit by simply placing a meter across the input to the op amp. This works because the current of the op amp varies approximately one decade for every 60 mV of change.—"RCA Solid State Databook—Integrated Circuits for Linear Applications." Courtesy of Harris Semiconductor.

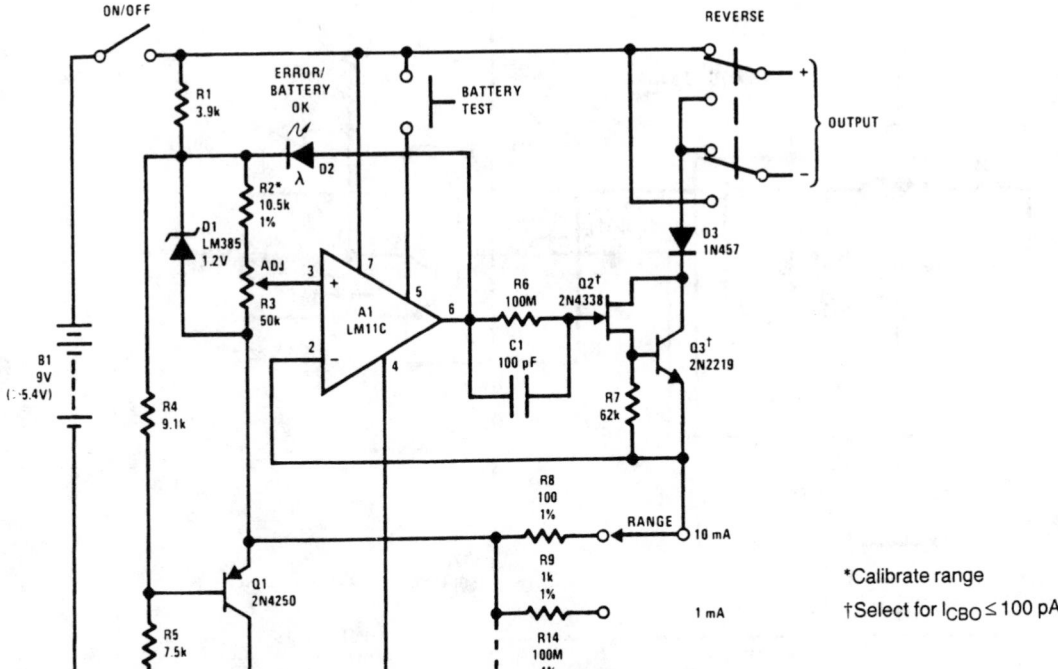

10-Microampere to 10-Milliampere Current Source.   This circuit represents
the classical op amp voltage to current conversion. The output resistance is
determined by both the matching and the value of the feedback resistors. This
circuit is unipolar in nature but can easily be converted to bipolar output if
desired. Typically, the top unipolar performance can be obtained more simply.
Output accuracy is determined by the 1N457 reference diode. The addition of
the 2N4250 P-N-P transistor at the input to this op amp ensures that the inputs
are kept within the common-mode range as battery voltage levels diminish. A
9-V DC power source is provided, but this circuit will work perfectly down to a
potential of 6 V DC and a bit less. An LED power supply indicator is provided
to make certain that the source is operating correctly. As long as the LED is lit,
there is sufficient battery voltage for the circuit to operate normally. A
momentary switch that simply causes the output to move toward a positive
value is provided for battery testing. All in all, this is a very complete circuit
that is relatively simple to construct.—"Linear Databook I." Courtesy of
National Semiconductor Corporation, Santa Clara, California.

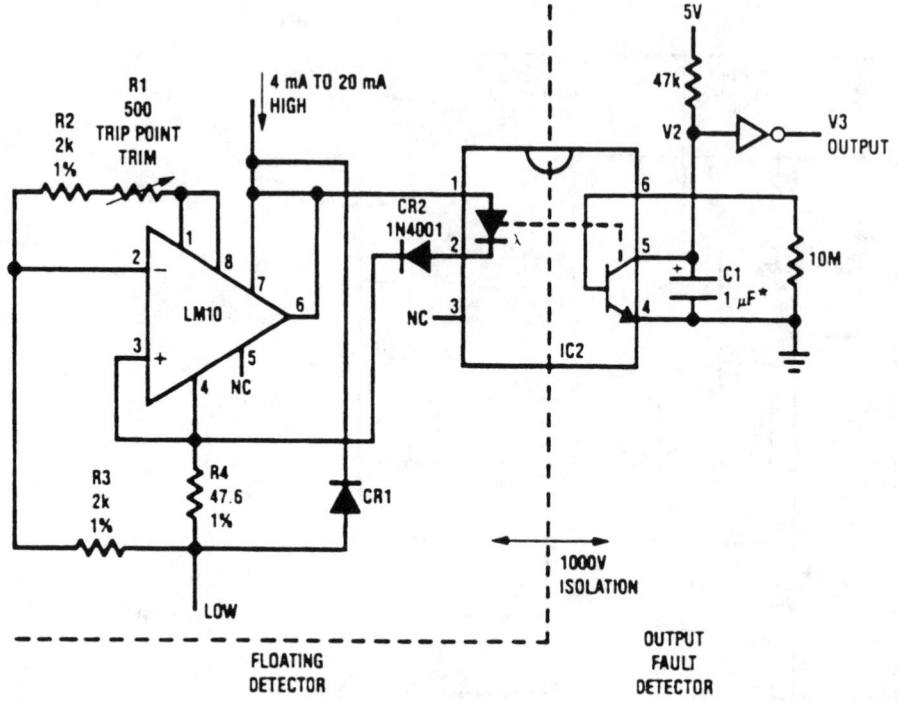

CR1 = 1N4001, optional, in case of signal reversal

LM10 = NSC LM10CLN or LM10CLH amp/reference

IC2 = 4N28 or similar, optoisolator

V2 = Normally low; high signifies fault (I < 3.7 mA)

V3 = Normally high; low signifies fault (I < 3.7 mA) (buffered output)

*C1 = 1 μF optional, to avoid false output when large AC current is superimposed on 4.0 mA.
    Disconnect this capacitor when using with circuit of *Figure 2*.

⊳o = 1/6 MM74C04 or similar, CMOS inverter

Current Loop Fault Detector.   This circuit can detect a loss or degradation of
signal below 4 mA with simplicity and low cost. It uses the LM10 op amp/
voltage reference chip. As long as the loop current is above 4 mA, the output is
off. The 4 to 20 mA will flow through the LED in the optoisolator and provide a
low output at its pin 5. When the loop current falls below about 3.7 mA, input
at pin 2 will rise and cause the output at pin 6 to fall, sealing all of the current
from the LED.—"Linear Applications Databook." Courtesy of National
Semiconductor Corporation, Santa Clara, California.

Chapter 6

Stereo/Audio Circuits

Circuits in this section address operations that are most generally associated with stereo applications and/or that portion of the audio spectrum inhabited by public address systems, communications equipment, and so on. It is in these areas that the multioperational amplifier chips are highly useful. The high inherent balance characteristics of a dual operational amplifier IC, for instance, are very difficult to achieve using two discrete op amps. Some op amp ICs contain three, four, or even more op amps and are especially suited to multichannel applications.

Op amps are also highly useful in tone control circuitry for bringing out the bass and treble responses in a complex signal. Several of these circuits are presented as well as an op amp filter for stereo applications. (Note: More op amp filter circuits are found in Chapter 9, which is devoted to this circuit class.)

Also found in this chapter are a number of power circuits that enhance the normally low level output of op amps with a boost to as much as 90 W. Naturally, this power increase is accomplished using discrete components, but the interface is quite simple, resulting in a good transition into the "power" band of the op amp spectrum.

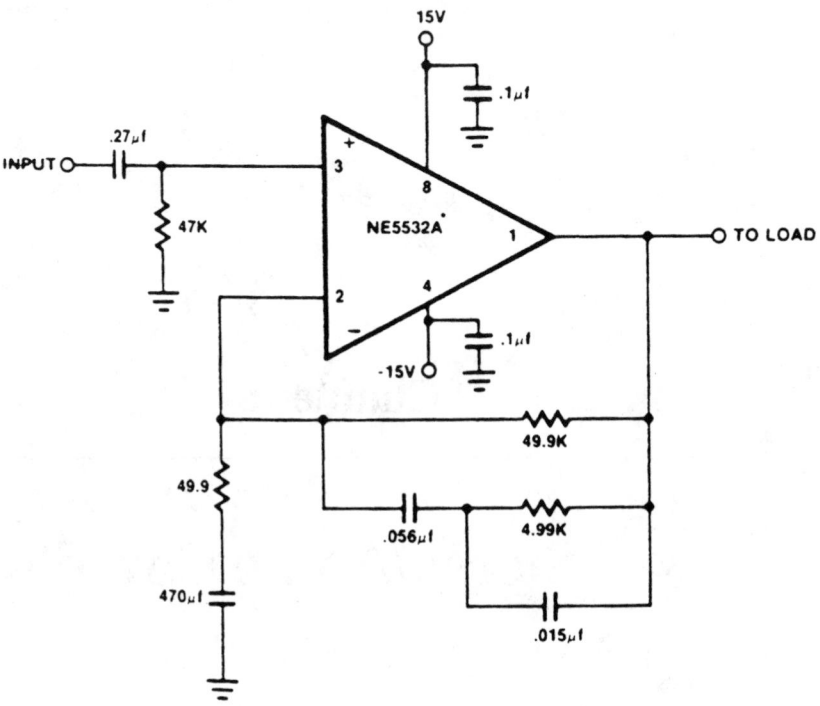

* 1/2 OF DUAL OPERATIONAL AMPLIFIER

Phonograph Preamplifier.   The Signetics NE5532 series of low noise op amps is a good choice for designing demanding input channels for critical audio systems. This circuit should serve the professional audio designer well. Obviously component selection, quality, and placement are critical in order to achieve the ultra low noise performance desired. All resistors should be 1% metal film types. This preamplifier offers excellent response between 31.5 Hz and 20 kHz. The response curve below 31.5 Hz is purposely suppressed to avoid turntable rumble that is so common with DC-coupled amplifier sections that pass frequencies in the 2 to 15 Hz range. This amplifier circuit meets the RIAA standard playback response curve.—"Linear Data Manual Volume 2: Industrial." Courtesy of Signetics Company, a division of North American Philips Corporation.

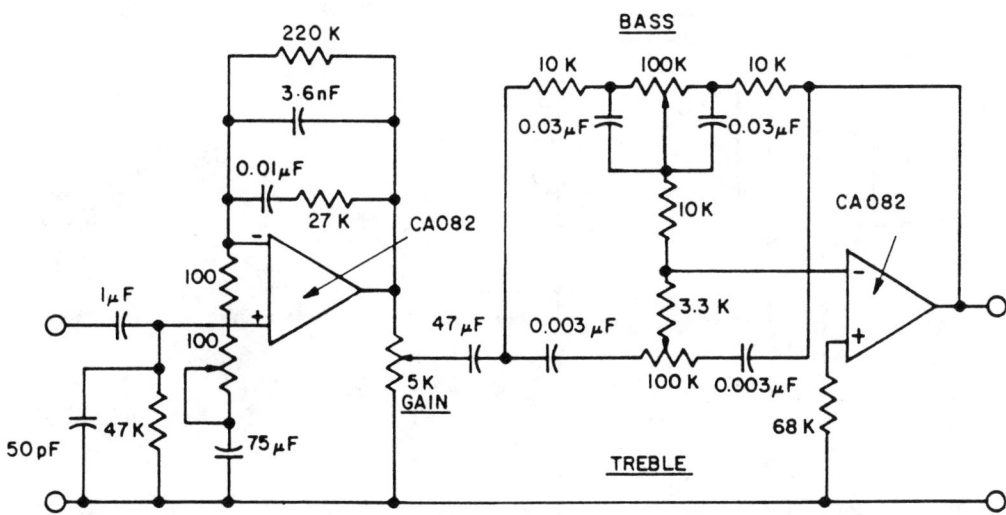

Preamplifier. This single-channel preamplifier uses a single RCA CA082 dual op amp in a design that also offers excellent bass/treble control. This series of op amps incorporates a MOS/FET input and composite bipolar/MOS output. Such a circuit can easily be placed in exiting audio equipment to add an additional channel. A typical power supply for this circuit would provide a bipolar 15-V DC output, however, lower values down to about 6 V DC will also suffice.—"RCA Solid State Databook—Integrated Circuits for Linear Applications." Courtesy of Harris Semiconductor.

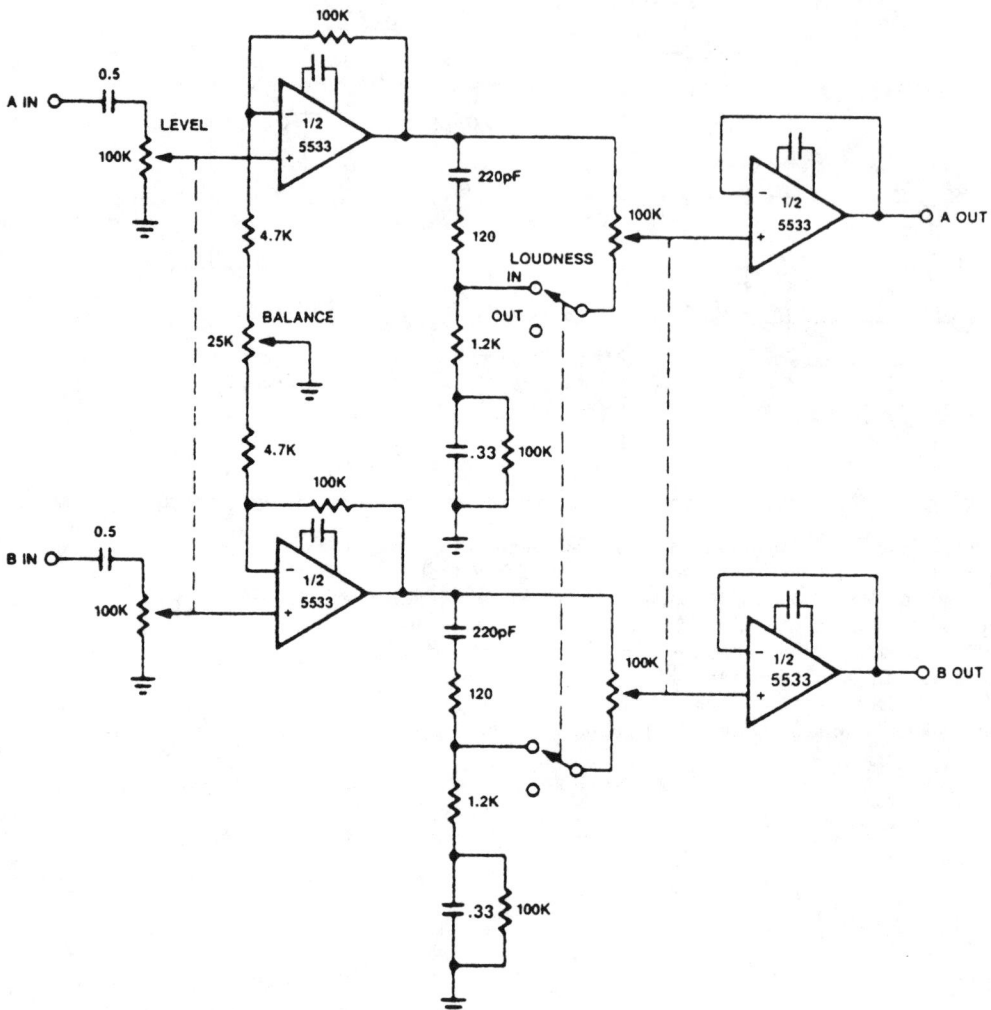

Balance Amplifier. This is a combination two-channel balance circuit coupled to a two-channel loudness control circuit and requires two Signetics NE5533 ultra low noise op amps. The first op amp allows the two channels to be perfectly balanced via the 25-kΩ potentiometer bridge. The dual output from this circuit feeds the second 5533. A switchable loudness control pads the output of the first stage through a dual RC network. This circuit should be built from the highest quality components in order to take advantage of the excellent low noise characteristics of this op amp.—"Linear Data Manual Volume 2: Industrial." Courtesy of Signetics Company, a division of North American Philips Corporation.

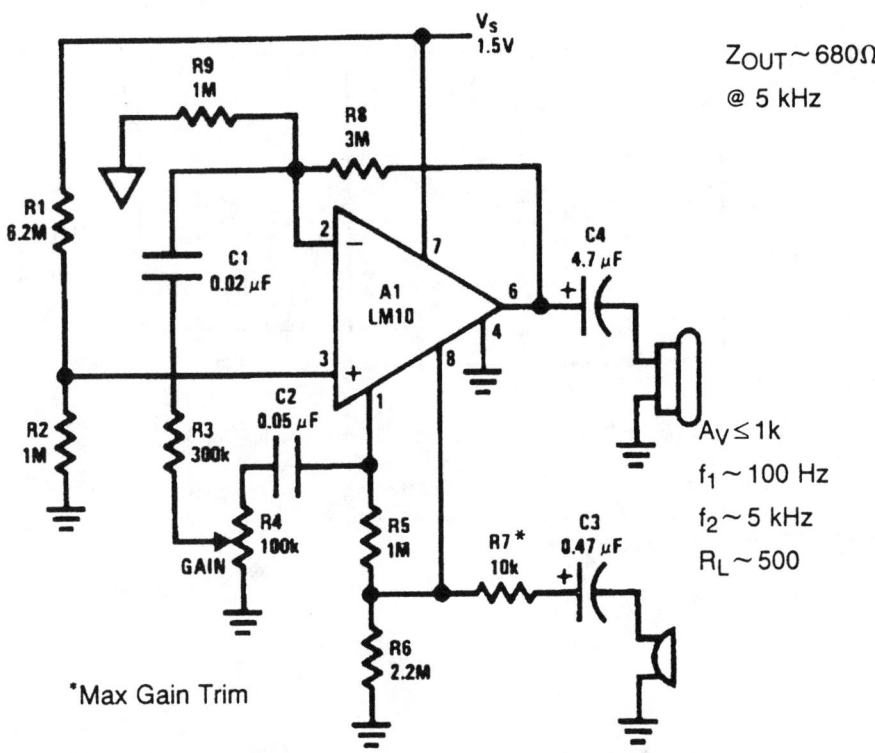

$Z_{OUT} \sim 680\Omega$
@ 5 kHz

$A_V \leq 1k$
$f_1 \sim 100$ Hz
$f_2 \sim 5$ kHz
$R_L \sim 500$

*Max Gain Trim

Microphone Amplifier. While the frequency response of the LM10 op amp is not ideal for audio purposes, it is adequate for construction of a simple microphone amplifier with a gain of approximately 60 dB and a bandwidth of about 5 kHz. In this usage, the voltage reference is used as the preamplifier with a gain of 100. Its output is fed through a 100K gain control to the op amp portion of the IC which produces a gain of 10. Input impedance is 10 kΩ and the output power is sufficient to drive a small earphone.—"Linear Databook I." Courtesy of National Semiconductor Corporation, Santa Clara, California.

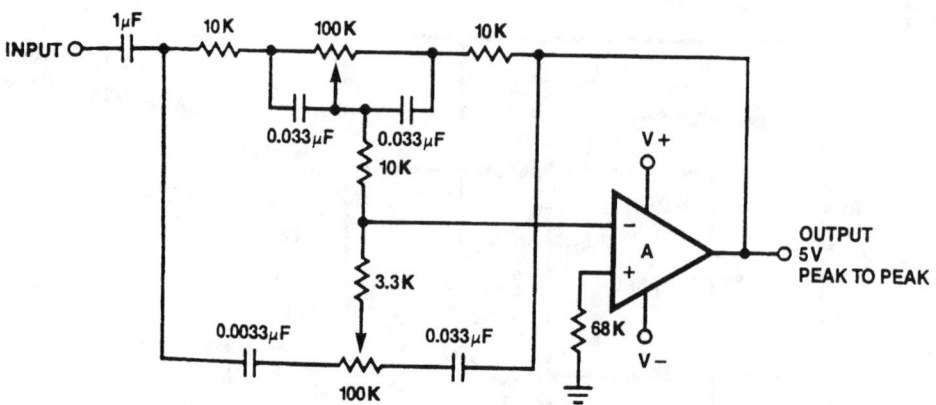

Tone Control. This op amp tone control provides about 20 dB of bass or treble boost or reduction as set by the 100-kΩ potentiometer. Turn-over frequency is 1 kHz. The op amp is a Signetics NE531 or NE301.—"Linear Data Manual Volume 2: Industrial." Courtesy of Signetics Company, a division of North American Philips Corporation.

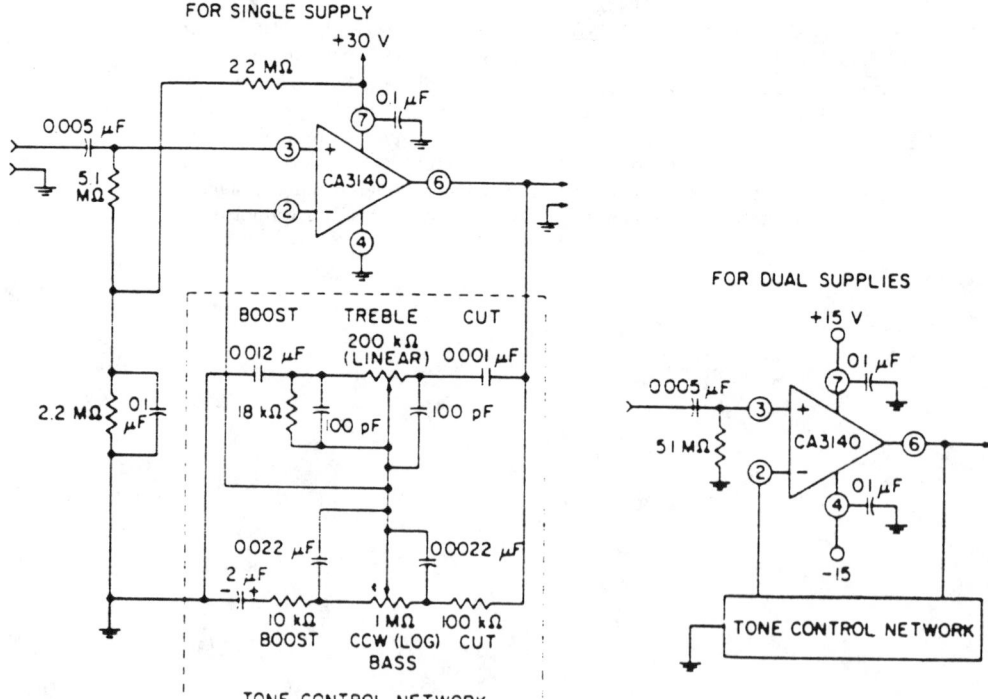

Baxandall Tone Control Circuit. This circuit provides unity gain at midband and uses standard linear taper potentiometers. The high input impedance of the RCA CA3140 op amp allows the effective use of low-cost, low-value capacitors. Bass/treble boost and cut are ±15 dB at 100 Hz and 10 kHz, respectively. Full output is available past 20 kHz and about −3 dB down at 70 kHz. This is an economical circuit that may be configured to operate from single or dual polarity supplies.— "RCA Solid State Databook—Integrated Circuits for Linear Applications." Courtesy of Harris Semiconductor.

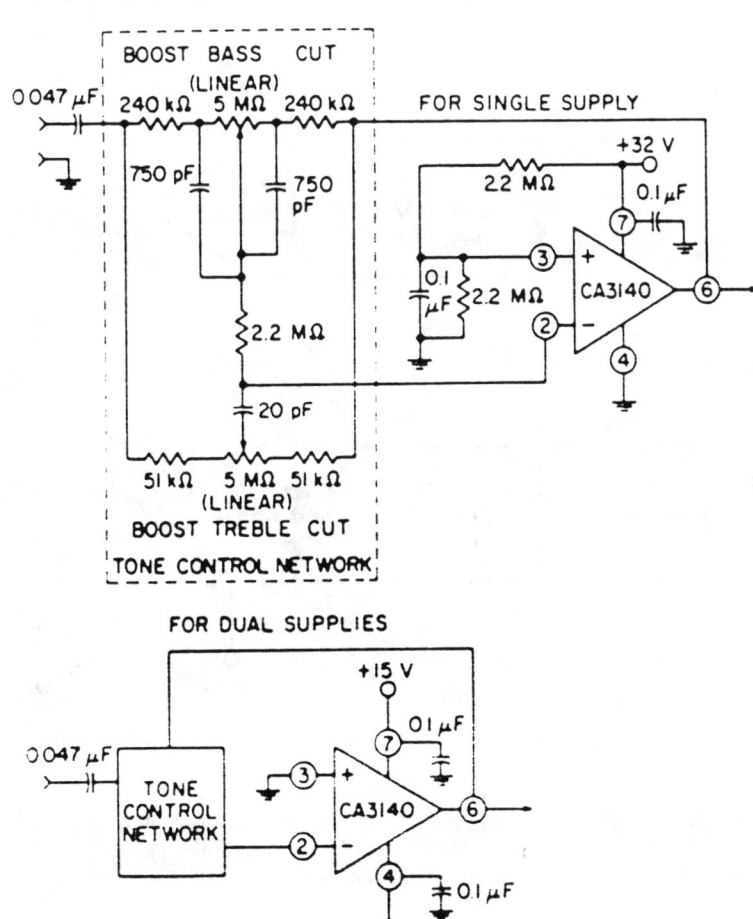

◀ **Wideband Tone Control Circuit.** This circuit provides a 20-dB midband gain and offers flat response up to about 20 kHz. For this boost level and cut, the input loading of this circuit is roughly equivalent to the value of measured resistance from pin 3 of the op amp to ground.—"RCA Solid State Databook— Integrated Circuits for Linear Applications." Courtesy of Harris Semiconductor.

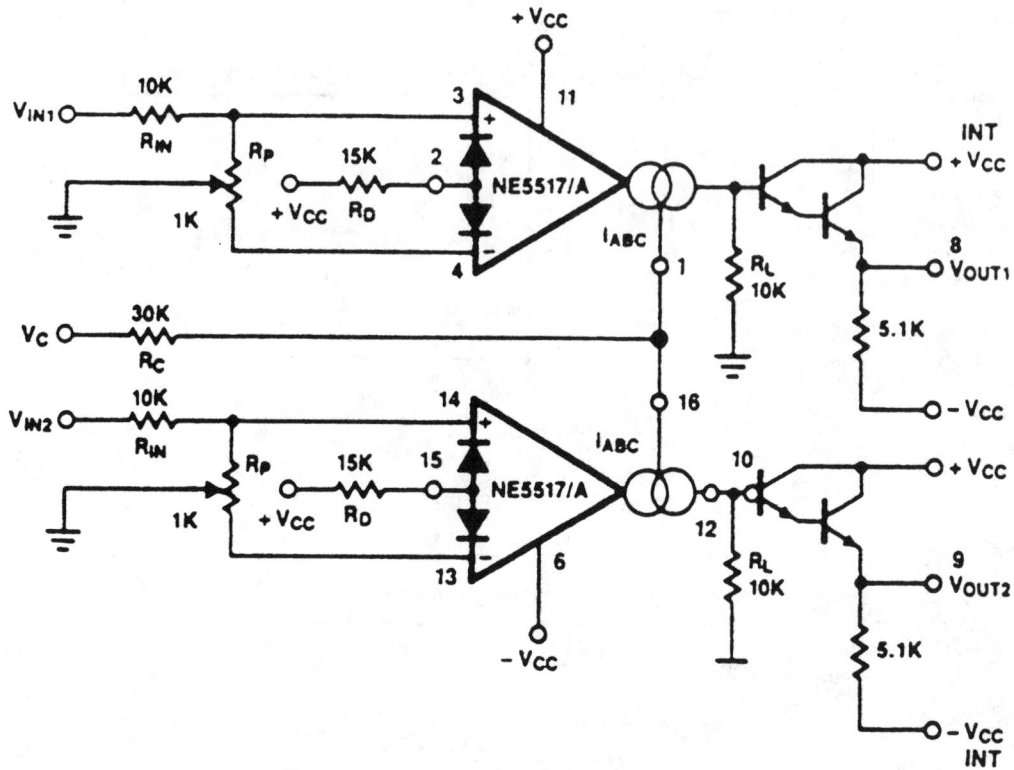

Gain-Controlled Stereo Preamplifier. This circuit offers excellent tracking as is usually the case with dual or quad op amps.—"Linear Data Manual Volume 2: Industrial." Courtesy of Signetics Company, a division of North American Philips Corporation.

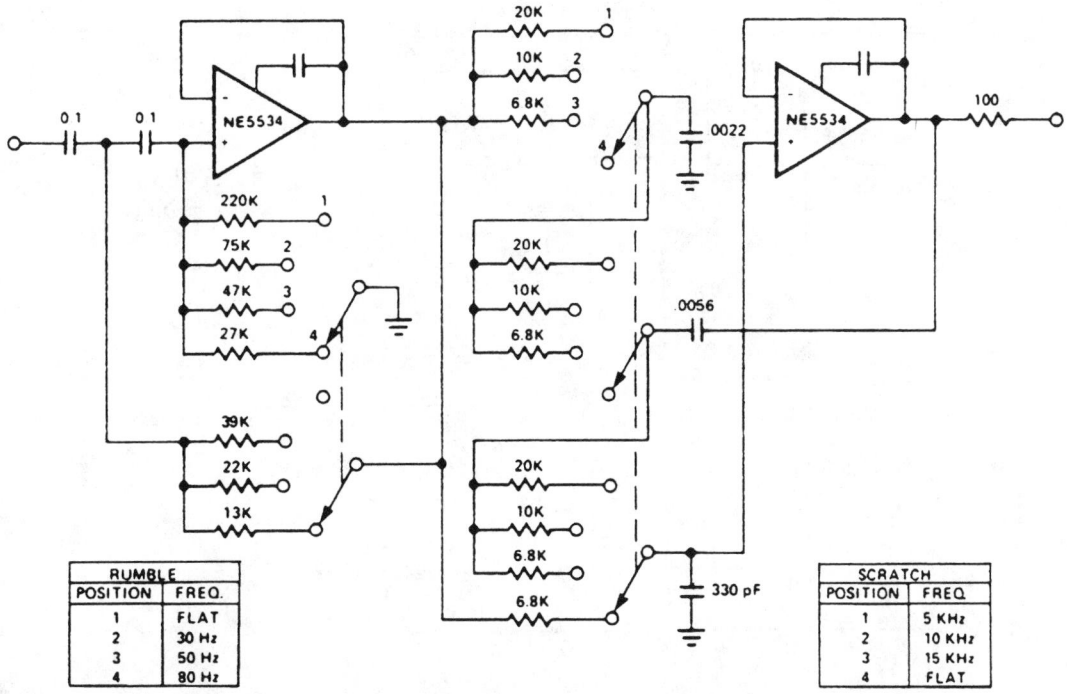

RUMBLE	
POSITION	FREQ.
1	FLAT
2	30 Hz
3	50 Hz
4	80 Hz

SCRATCH	
POSITION	FREQ.
1	5 KHz
2	10 KHz
3	15 KHz
4	FLAT

Rumble/Scratch Filter.   This rumble/scratch filter is designed to follow the first amplifier stage of a sophisticated stereo system using a turntable. This is a two-pole Butterworth approach that features switchable break points. This circuit offers a wide variation of filtering from very sharp to none at all. Make certain that high-quality switches are used. The cheap varieties may render this circuit more of a hindrance than an asset.—"Linear Data Manual Volume 2: Industrial." Courtesy of Signetics Company, a division of North American Philips Corporation.

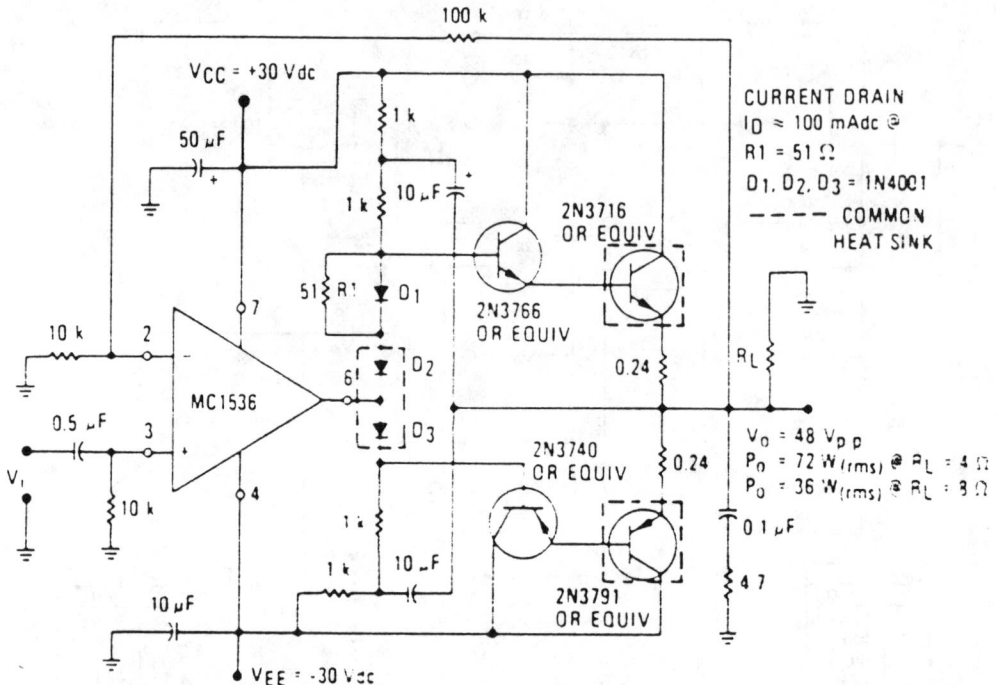

Audio Amplifier. This circuit uses the Motorola high-voltage op amp and operates from a bipolar 30-V DC supply to output up to 72 W (rms) to a 4-Ω load or half that amount to an 8-Ω load.—"Motorola Linear and Interface Integrated Circuits." Courtesy of Motorola Semiconductor Products, Motorola, Inc.

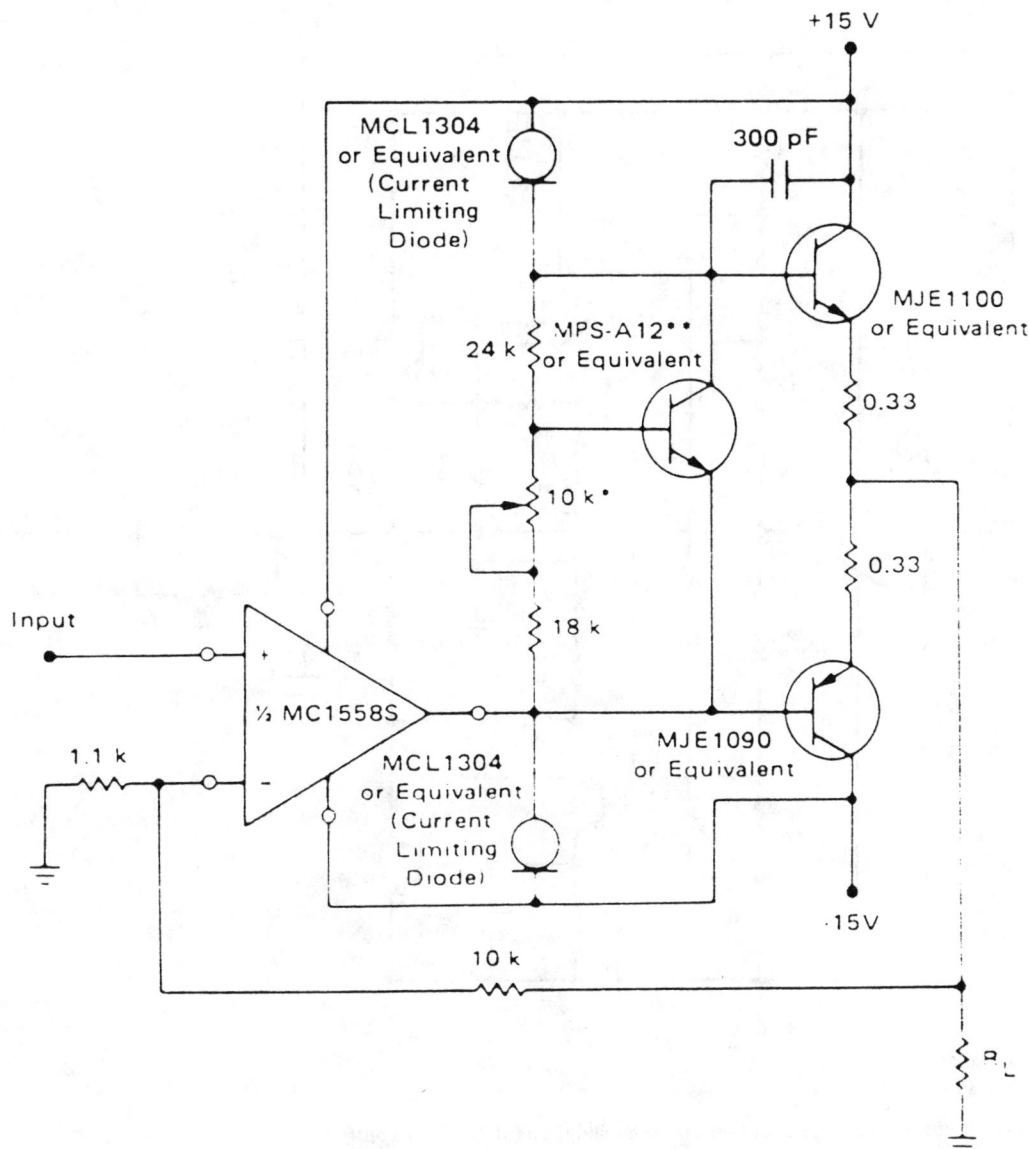

12-Watt Wideband Power Amplifier. This hybrid circuit delivers slightly over 12 W into a 4-Ω load with typically less than 1% THD to 100 kHz. The 10-kΩ potentiometer is used for bias current adjustment to eliminate crossover distortion.—"Motorola Linear and Interface Integrated Circuits." Courtesy of Motorola Semiconductor Products, Motorola, Inc.

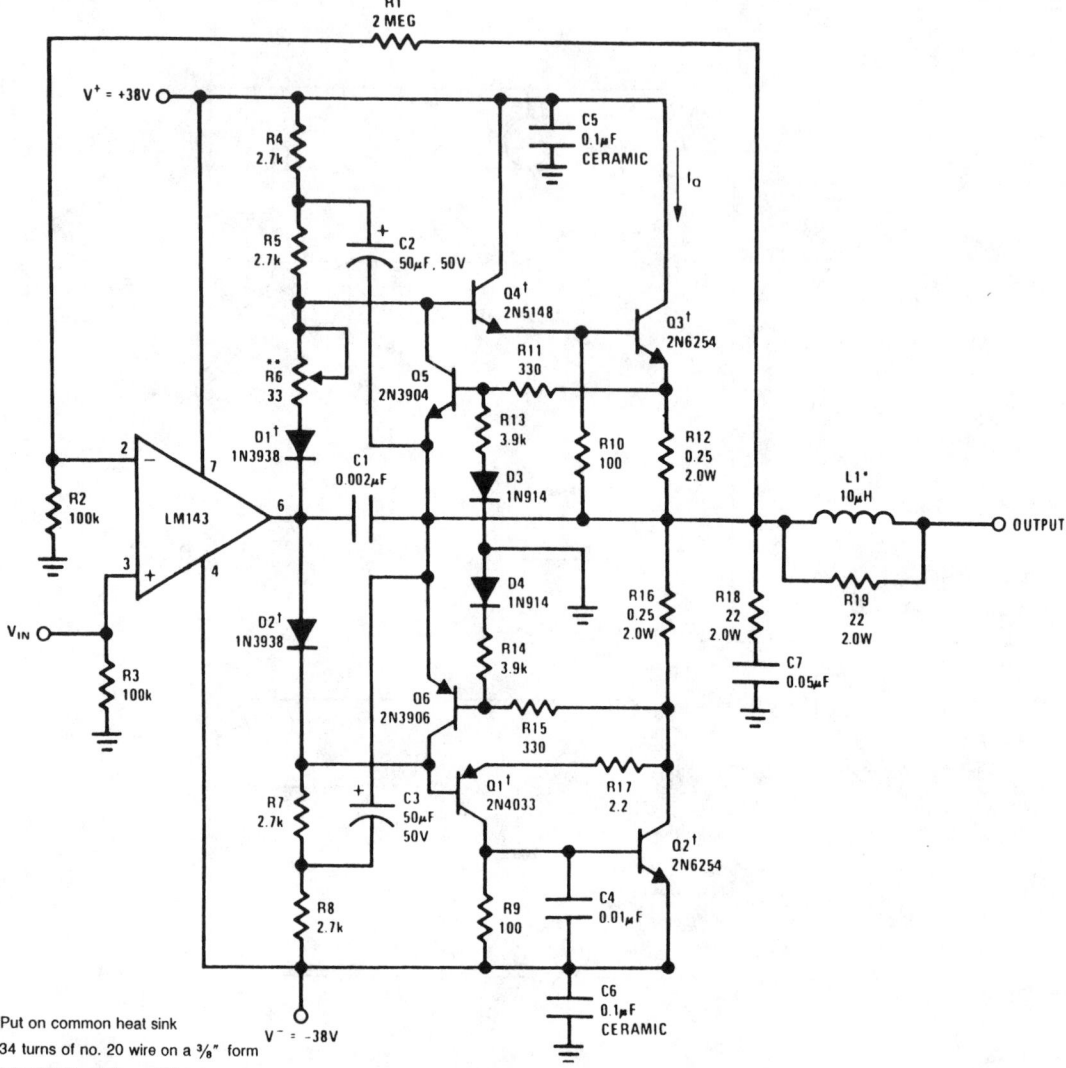

†Put on common heat sink
*34 turns of no. 20 wire on a ⅜" form
**Adjust R6 to set I_Q = 100 mA

90-Watt Power Amplifier. This circuit is capable of a full 90-W rms output to
a 4-Ω load or 70 W to 8 Ω. It is fully protected from short circuit at the ouput
and from overloads. It offers excellent harmonic distortion characteristics for
an amplifier of this power, typically charting in at 0.1% at 1000 Hz. The power
is derived from the N-P-N transistor output stage. The output of the op amp
drives a quasi-complementary output stage made up of Q1 through Q4. This
amplifier is biased into class AB operation by using a resistor string comprised
of R4, R5, R7, and R8 to set the voltage drops across R6, D1, and D2. The output
inductor is hand wound from number 20 enamel clad wire. It consists of 38
close-wound turns on a ⅜-in. diameter form. This circuit is not too difficult
from a construction standpoint, but the heatsinking requirements will add to
the mechanical complexity. A Thermalloy 6006B heat sink is recommended
for common mounting of all power components.—"Linear Databook I."
Courtesy of National Semiconductor Corporation, Santa Clara, California.

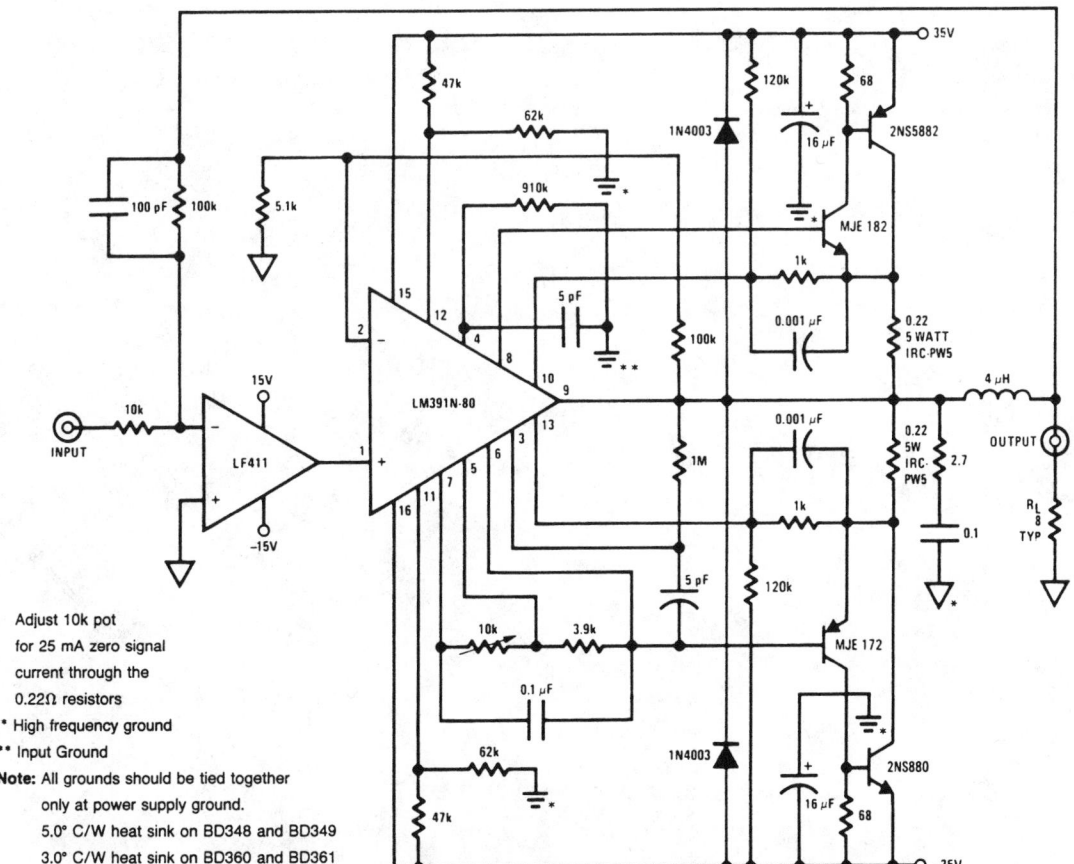

High-Current Op Amp Booster.   This circuit has the capability of driving 3 A
of current into an 8-Ω load at 25 V peak. The LF411 op amp has an LM391-80
driver chip placed in its feedback loop. The 5 pF capacitor at pin 3 of the
driver chip sets the booster bandwidth to in excess of 250 kHz. The 10-kΩ
potentiometer is used to set the zero signal current at the output stage.—
"Linear Applications Databook." Courtesy of National Semiconductor
Corporation, Santa Clara, California.

Chapter 7

Timer/Control Circuits

While operational amplifiers are not immediately thought of as timer/control devices, they are utilized quite often, especially as timing references, in conjunction with other types of ICs more closely associated with this application. In this chapter, timing circuits of various duration are featured along with the usual assortment of light flashers.

IC operational amplifiers are also found more and more in control circuits, many of which are designed for moderate- to high-power applications when used in conjunction with discrete power components. However, there are a few operational amplifiers that can directly offer moderate-power motor control capabilities, up to about 0.5 A. Several circuits that address this type of application are also presented in this chapter.

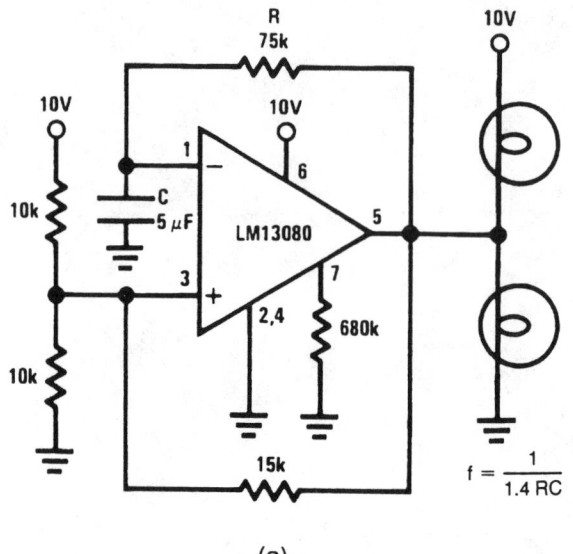

(a)

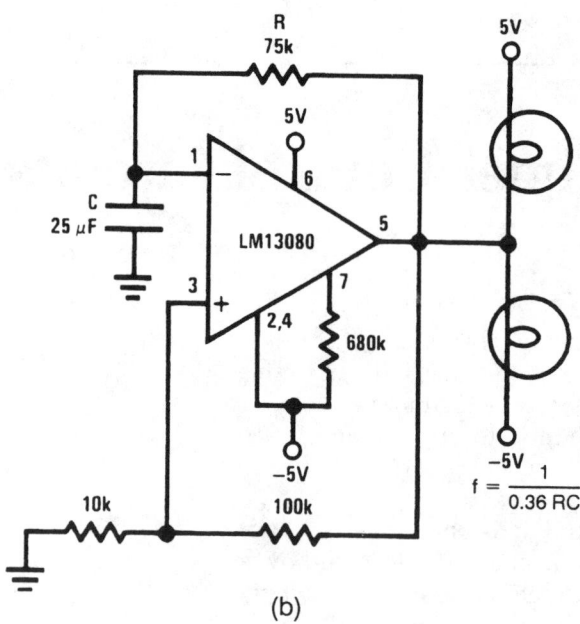

(b)

Lamp Flasher. The LM13080 programmable power op amp is an excellent device to use for lamp flashing operations. The flasher is a simple square wave oscillator with a low-frequency output. This output waveform is used to drive small incandescent lamps in these examples, however LEDs, relays, and so on may also be directly driven in the same fashion. The first example (a) uses a 9–10-V DC single polarity supply while the second (b) incorporates a dual polarity supply at 5 V DC. Flash-rate-determining components are C and R. As their values increase, flash rate decreases. Try substituting a 100-kΩ variable resistor for R to convert to a variable flash-rate design.—"Linear Databook I." Courtesy of National Semiconductor Corporation, Santa Clara, California.

100-Second Timer. This 100-sec timing circuit can be modified to produce a wide range of timing intervals. C is a 1-µF unit, while R1 and R2 are rated at 1000 Ω. R3 and R4 are 144-MΩ 0.5-W carbon resistors. R5 is 2000 Ω and R6 is a 20-kΩ unit.—"Motorola Linear and Interface Integrated Circuits." Courtesy of Motorola Semiconductor Products, Motorola, Inc.

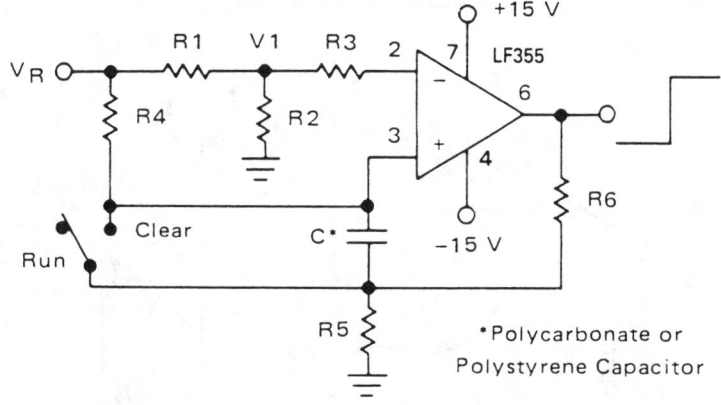

RC Timer. On a negative-going transient at the input, a negative pulse at point C will turn the op amp on, and the output will go from low to high. The time constant is determined by the values of R1, R2, and R3, and by C1 and determines the time delay before the output switches back to a low state. Upon time out, the op amp is turned off and the output is pulled to the resting (low) state by the load. This condition is independent of the interval required for the input to return to a high level.—"RCA Solid State Databook—Integrated Circuits for Linear Applications." Courtesy of Harris Semiconductor.

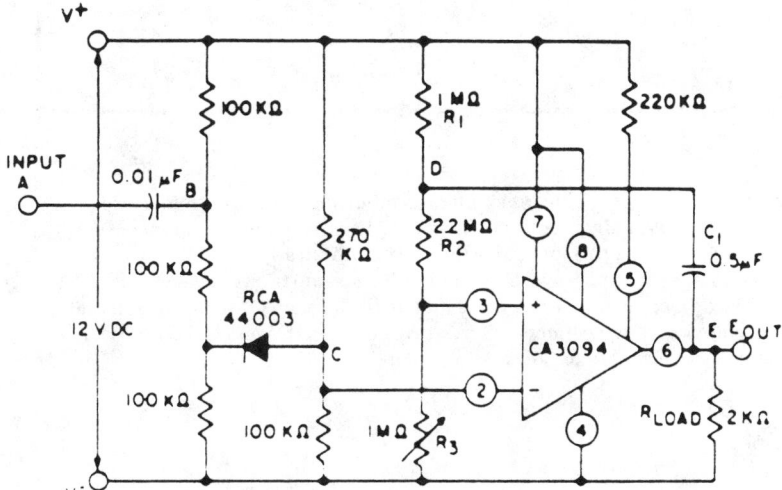

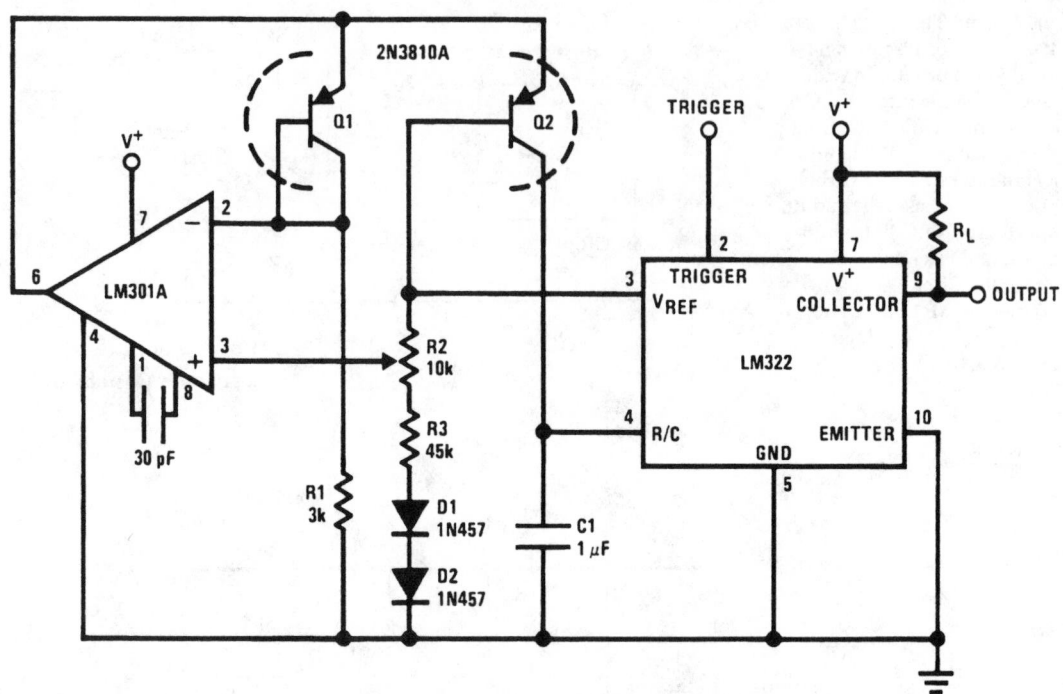

Wide-Range Timer. This timer offers an adjustment range of from 2 msec to 2000 sec with a single control. An LM322 is used as a precision timer in conjunction with a current source that is logarithmically controlled from a potentiometer. The noninverting input of the op amp is a voltage reference. R1 and the op amp set up a 1-mA constant current through Q1 using the internal 3-V reference from the timer.—"Linear Applications Databook." Courtesy of National Semiconductor Corporation, Santa Clara, California.

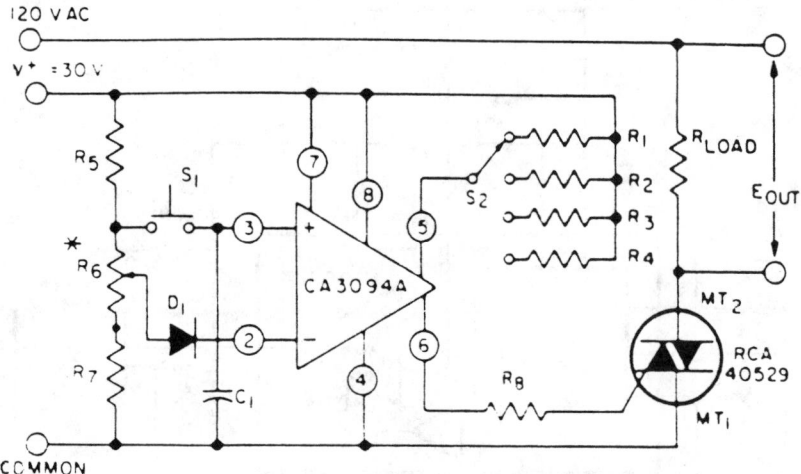

Presetable Analog Timer. Resistors R1 through R4 determine the time lag before the thyristor is triggered. Their values are 510 kΩ, 5.1 MΩ, 22 MΩ and 44 MΩ respectively for intervals of 3, 30, 120, and 240 min. R5 and R8 are 1.5 kΩ and R7 is 5.1 kΩ. R6 has a nominal value of 50 kΩ. This potentiometer may be replaced with a fixed resistor after the circuit is adjusted. C1 is a 0.5-μF value and D1 is a 1N914 or equivalent.—"RCA Solid State Databook—Integrated Circuits for Linear Applications." Courtesy of Harris Semiconductor.

Motor Speed Control. It is rare to find an op amp that can serve as a complete motor speed control without resorting to the use of power transistors. The LM13080 is one that can be used to control the speed of small DC motors that require less than 0.5 A starting current. Additionally, this is a simple circuit that can be assembled from "junk box" parts. The two diodes can be any 1 A low-voltage types. R3 and D2 set up a reference voltage that is applied to the noninverting input. The bias resistor R4 assures that D2 remains active. R1/R2 is actually a single, linear taper potentiometer that acts like two resistors. This is the motor speed control. Adjust R3 for the smoothest operation of the motor at all positions of R1/R2. Once proper operation is obtained, R3 may be replaced by a fixed resistor of the same value as the measured resistance within the potentiometer.—"Linear Databook I." Courtesy of National Semiconductor Corporation, Santa Clara, California.

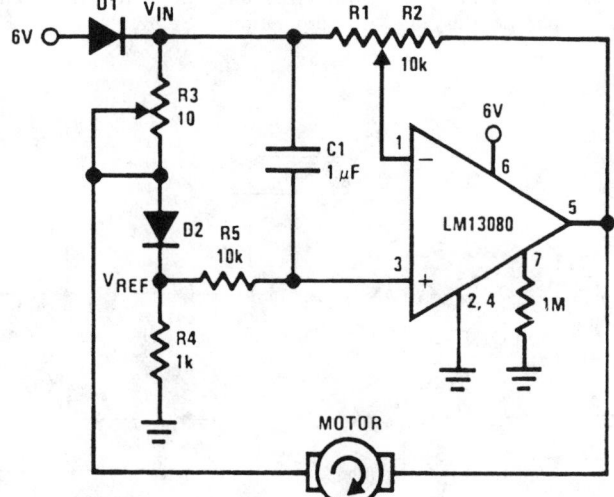

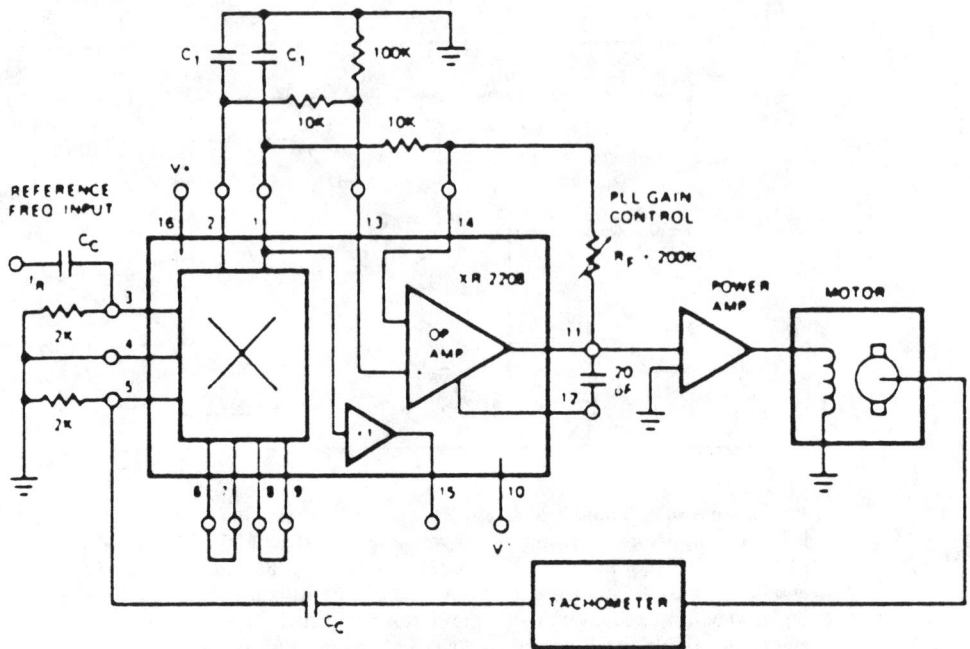

Motor Speed Control. This circuit uses the multiplier/op amp circuit combination of the EXAR XR-2208 IC as a motor speed control where the frequency of the motor is phase-locked to the input reference frequency. The multiplier section is the phase comparator. The resulting error voltage across pins 1 and 2 is filtered by C1 (low-pass) and amplified by the op amp.—"EXAR Databook." Courtesy of EXAR Corporation.

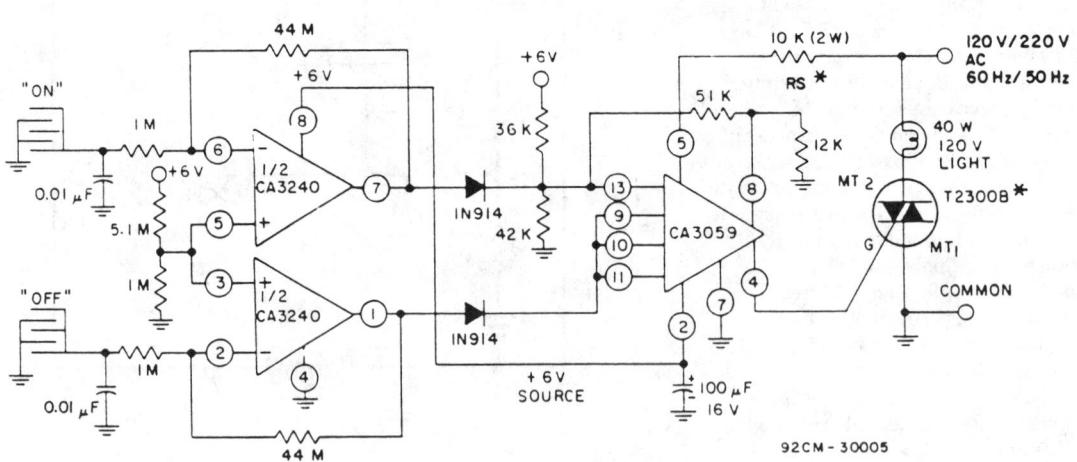

*AT 220 V OPERATION, TRIAC SHOULD BE T2300D,
  RS= 18 K, 5 W

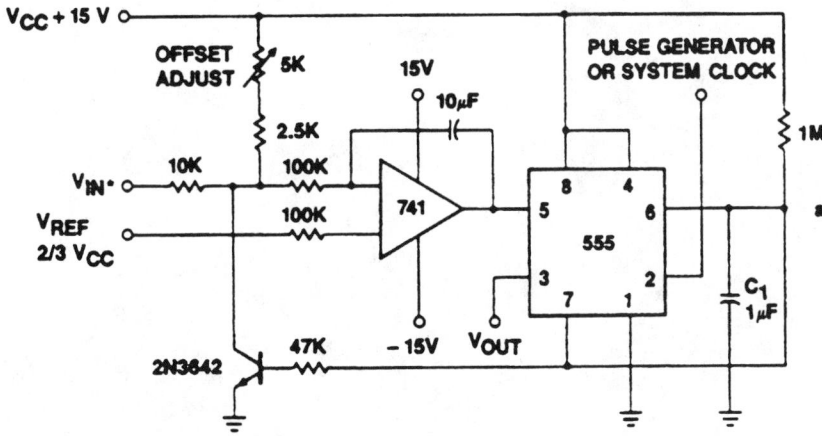

Voltage-to-Pulse Duration Converter. This circuit incorporates a 741 op amp and a 555 timer to offer voltage-to-pulse duration conversions with an accuracy factor of better than 1%. The output signals still retain the original frequency, independent of the input voltage.—"Linear Data Manual Volume 2: Industrial." Courtesy of Signetics Company, a division of North American Philips Corporation.

◀ **On/Off Touch Switch.** This circuit uses two different ICs in a configuration that will sense very small currents floating between two contact points. These points are the touch switches that are finger activated. These "switches" are most often square "coils" etched on a small piece of circuit board. When the *on* plate is touched, current flows between the two halves of the grid causing a positive shift in output voltage of the CA3240. These positive transitions are fed to the CA3059 which is used as a latching circuit and triac driver. When the *off* grid is touched, the triac is turned off and held off by the same type of function occurring within the circuitry associated with the second op amp within the CA3240 package.—"RCA Solid State Databook—Integrated Circuits for Linear Applications." Courtesy of Harris Semiconductor.

Chapter 8

Power Supply Circuits

The operational amplifier has assumed a very prominent role in the construction of sophisticated DC power supply circuitry. In most instances the op amp does not form the core of such circuits. Rather it is an adjunct component that allows a simpler circuit to be better controlled, thus offering improved range and/or regulation characteristics. For instance, adding an inexpensive op amp to a simple three terminal fixed regulator circuit will allow that circuit to offer a variable output and to provide output voltage values that are higher than those for which the regulator was initially designed. All of this improvement and control is obtained while still retaining the same regulation characteristics of the simpler circuit.

Op amps are also found in power supplies that produce relatively high-voltage outputs with excellent regulation, in constant-current/constant-voltage sources and in tracking regulators. A DC-to-DC power supply is also included in this chapter which uses the op amp as the reference source in its switching circuitry.

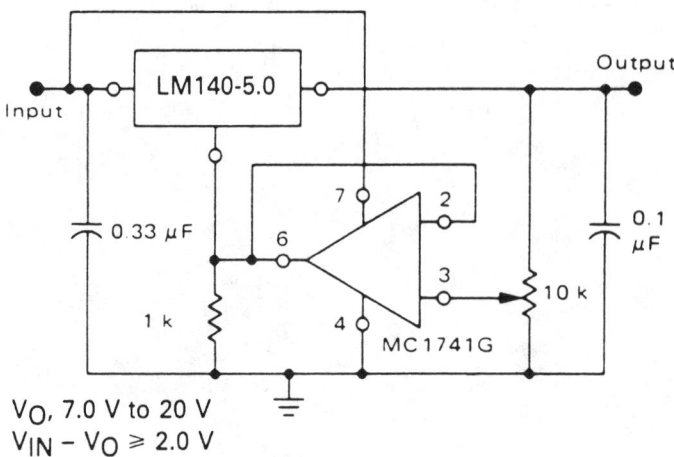

V_O, 7.0 V to 20 V

$V_{IN} - V_O \geq 2.0$ V

Adjustable Output Regulator.
The addition of an op amp to
this LM140 regulator allows for
the adjustment to higher or
intermediate values while
retaining the same regulation
characteristics. The minimum
voltage that may be obtained
from this circuit is 2.0 V greater
than the regulator voltage.—
"Motorola Linear and Interface
Integrated Circuits." Courtesy of
Motorola Semiconductor
Products, Motorola, Inc.

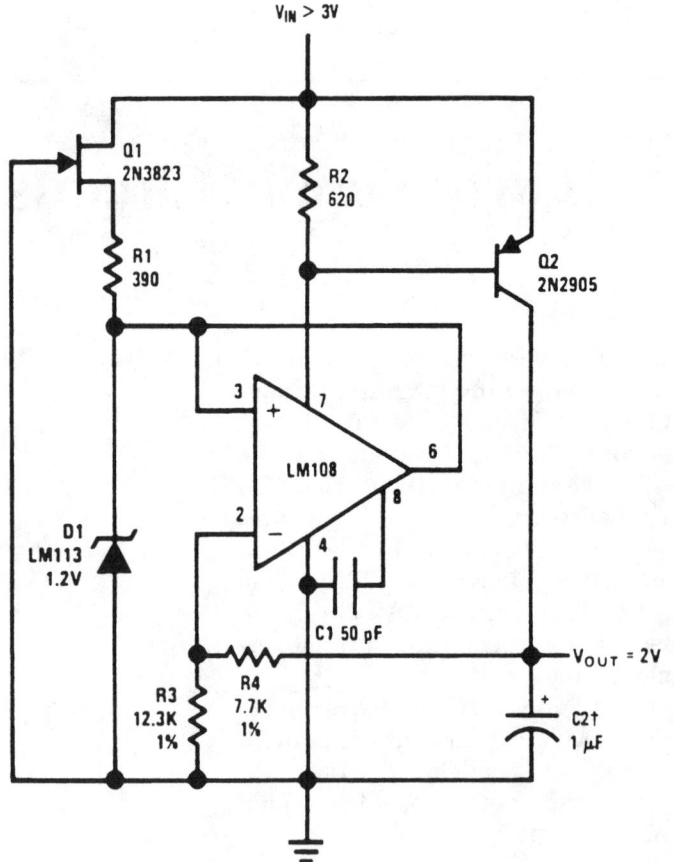

Low-Voltage Regulator. This
low-voltage regulator circuit
operates with an input of 3 V DC
(minimum) and outputs a
regulated 2 V DC. The heart of the
circuit is an LM113 1.2-V
temperature compensated shunt
regulator diode. It is driven by the
2N3823 FET current source. The
op amp compares a fraction of the
output voltage with the reference
value. Op amps other than the
LM108 can be used, but it is
important that other types have
low resting current demands. This
quiescent current flows through
R2 and tends to turn Q2 on.
Note: C2 should be a solid
tantalum type.—"Linear
Applications Databook." Courtesy
of National Semiconductor
Corporation, Santa Clara,
California.

High-Voltage Regulator.
This high-voltage
regulator circuit uses the
LM10 op amp as a voltage
reference source at the
base of Q2, which in turn
controls conductivity
through the output
transistor Q2. The input
supply must provide in
excess of 205 V DC which
will result in a regulated
output of 200 V DC. The
output voltage may be
varied by substituting
other values for R1 and R2
and by raising or lowering
the input voltage
accordingly. Output
transistor selection can be
changed to match voltage/
current needs.—"Linear
Databook I." Courtesy of
National Semiconductor
Corporation, Santa Clara,
California.

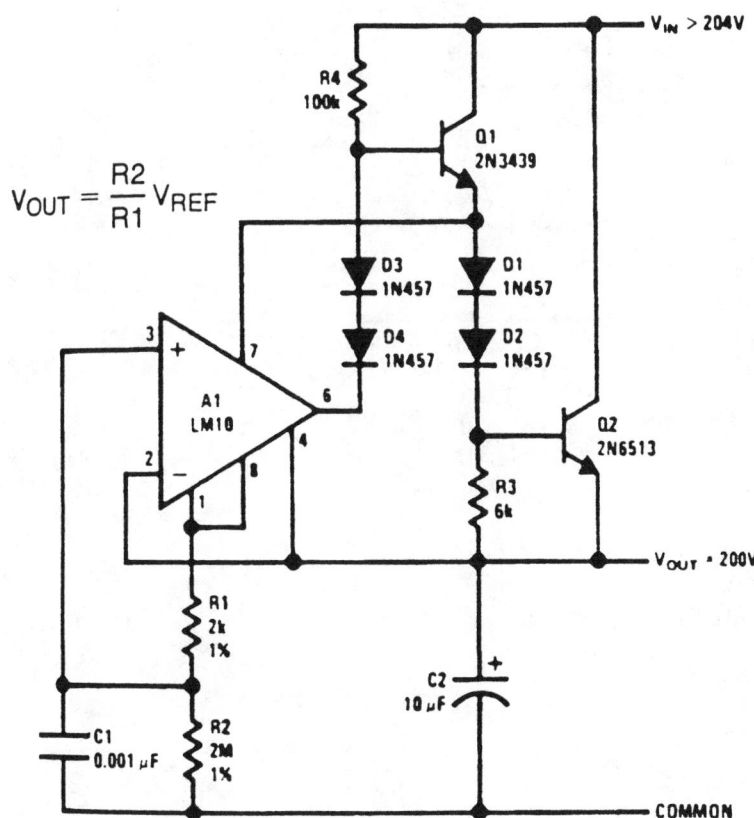

$$V_{OUT} = \frac{R2}{R1} V_{REF}$$

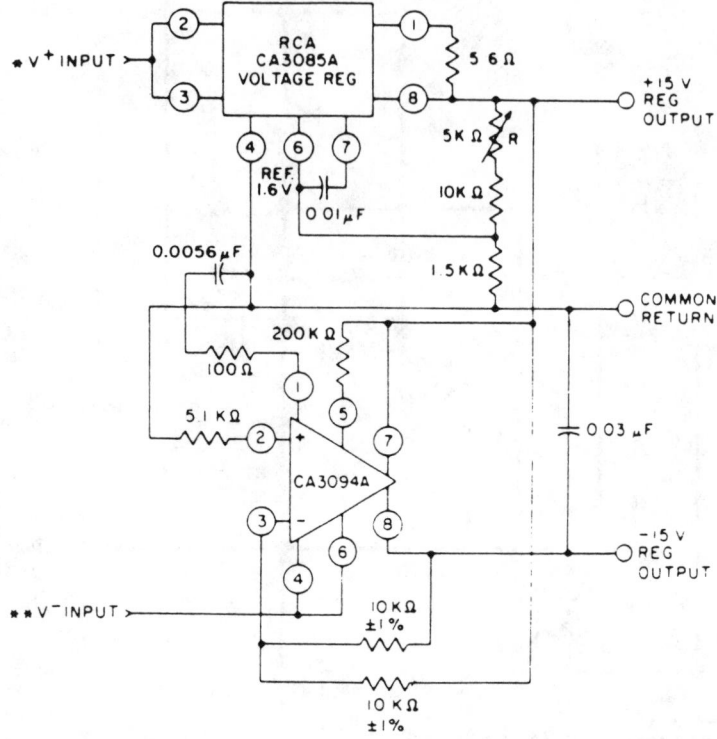

Dual-Voltage Tracking Regulator. Using the CA3094A programmable power switch/op amp, this circuit provides a bipolar 15-V DC output with the positive and negative supply always in sync regarding output voltage. Input potential should be in the 20- to 30-V range, although this value may drop to −16 V DC on the negative side while still maintaining a −15 V DC output. Maximum load for best regulation characteristics is 50 mA.—"RCA Solid State Databook—Integrated Circuits for Linear Applications." Courtesy of Harris Semiconductor.

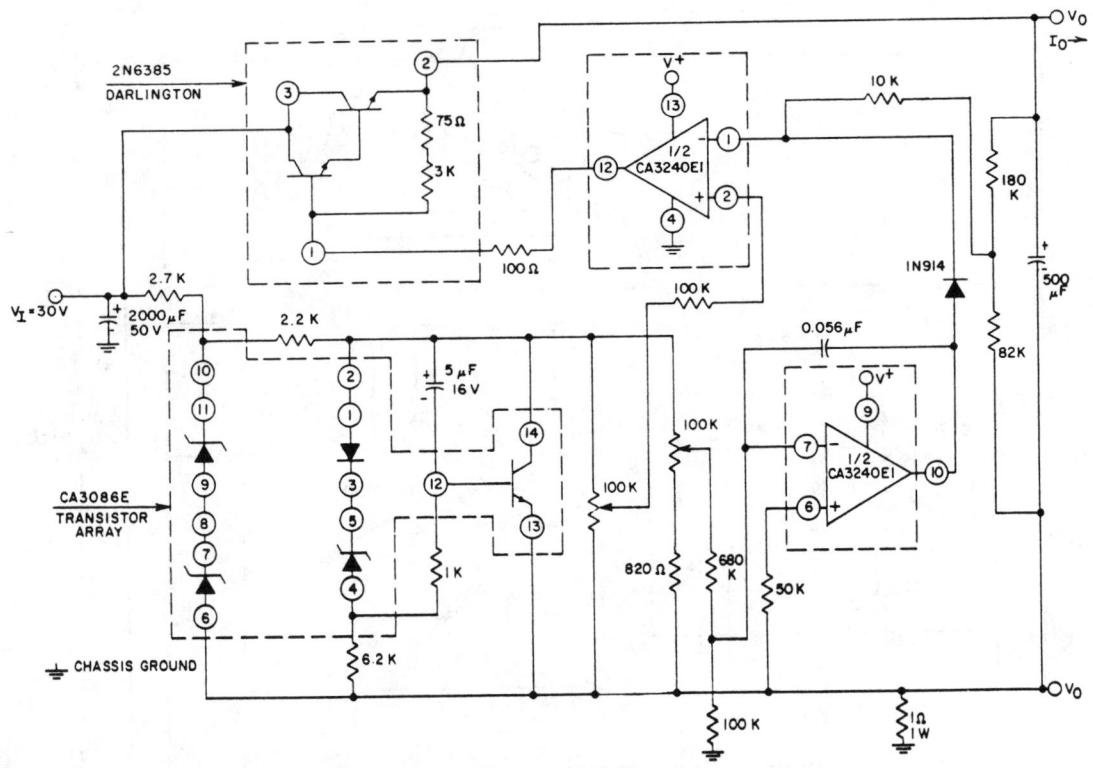

Constant-Voltage/Constant-Current Power Supply. A CA3240E MOSFET input op amp serves here as a voltage-error and current-sensing amplifier. Its common-mode voltage range includes ground, allowing the supply to adjust from 20 mV to 25 V without requiring an additional negative input voltage. The 2N6385 power Darlington is used as the pass element and may be required to dissipate as much as 40 W at current outputs in excess of 1 A.—"RCA Solid State Databook—Integrated Circuits for Linear Applications." Courtesy of Harris Semiconductor.

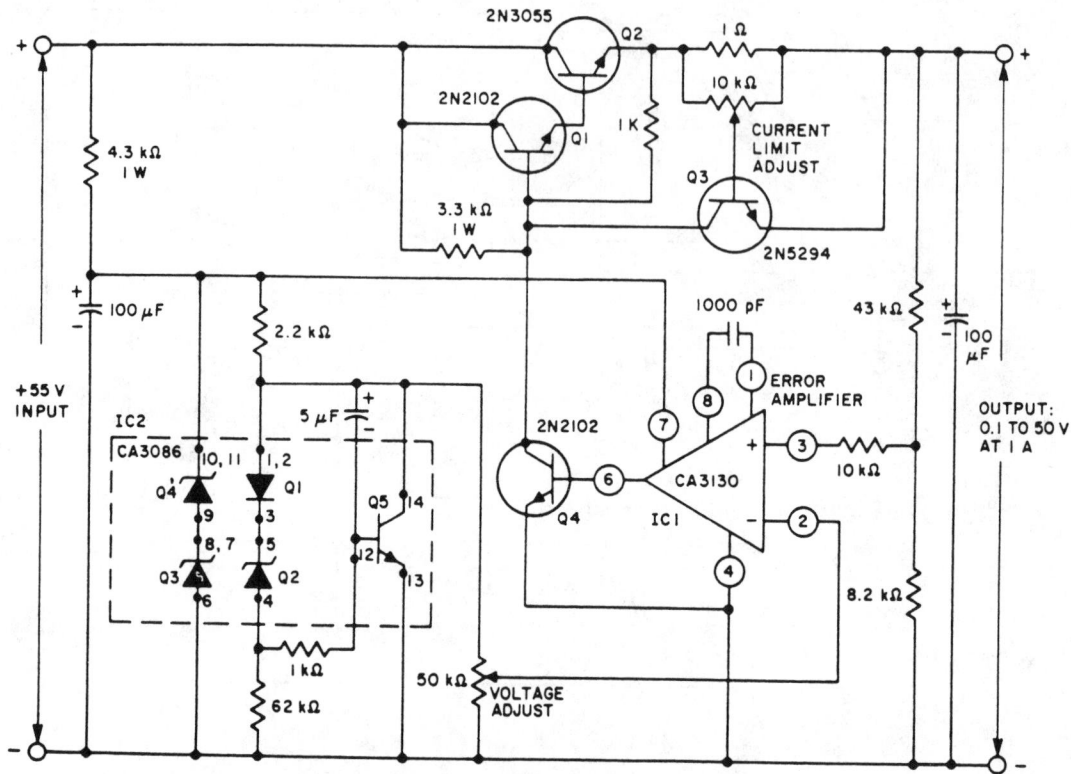

1-Ampere, Wide-Range Variable Regulator. This is an excellent circuit for constructing a very versatile bench power supply. It outputs a full 1 A over a very wide range of continuously variable voltages of from 0.1 V DC to 50 V DC. The DC power supply that drives this regulator must provide slightly more than 55 V DC for full 50-V regulated output, but lower input voltages may be used if outputs of this magnitude are not desired.—"RCA Solid State Databook—Integrated Circuits for Linear Applications." Courtesy of Harris Semiconductor.

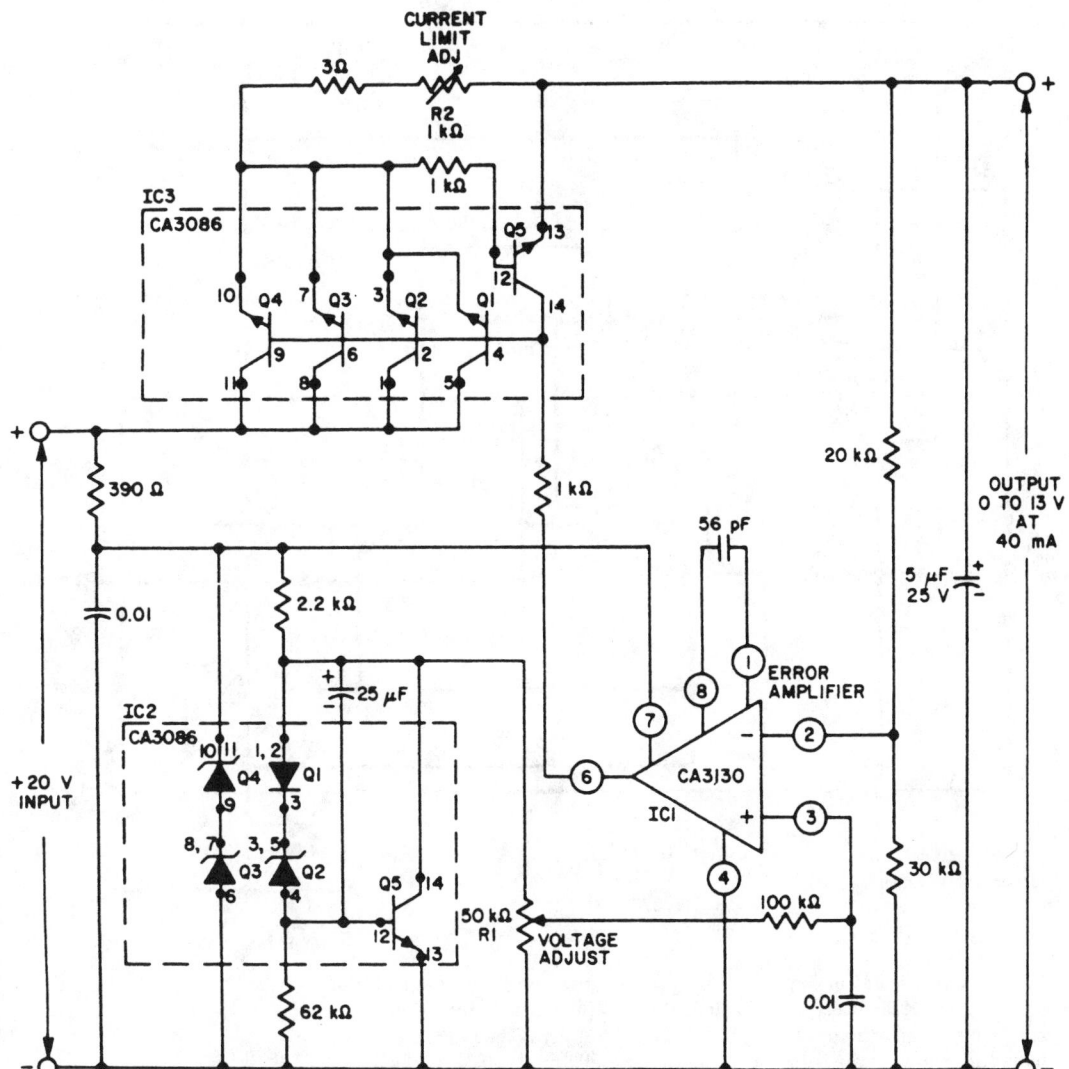

0–13-Volt DC Voltage Regulator. This circuit incorporates an op amp as an error amplifier in a 40-mA continuously variable power supply. It provides highly regulated output and uses two other ICs for compact, uncomplicated construction.—"RCA Solid State Databook—Integrated Circuits for Linear Applications." Courtesy of Harris Semiconductor.

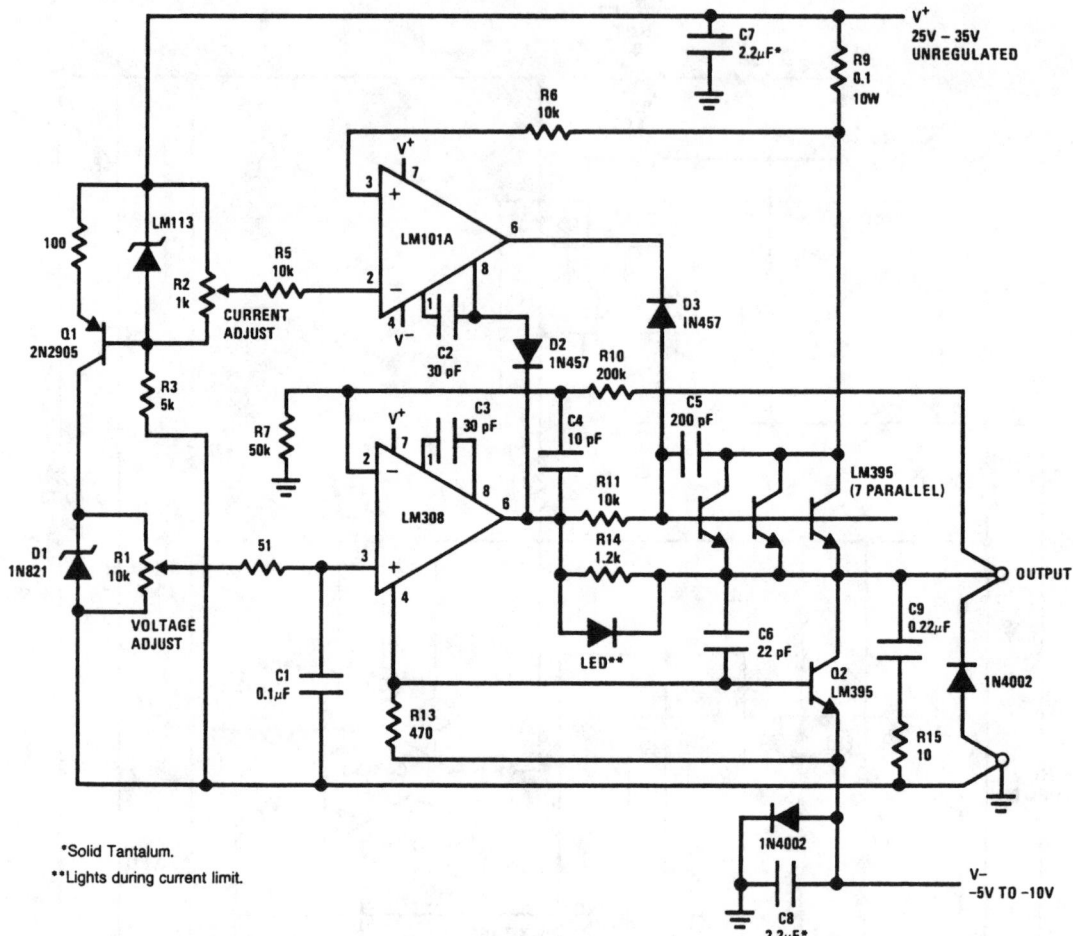

General Purpose Power Supply. This power supply will provide an output of up to 25 V DC at a maximum of 10 A. Both output voltage and current are adjustable down to zero. The key to this power supply is the LM395 monolithic power transistor (7 of them in parallel) which serves as the pass element. Obviously, heat sink design is critical.—"Linear Applications Databook." Courtesy of National Semiconductor Corporation, Santa Clara, California.

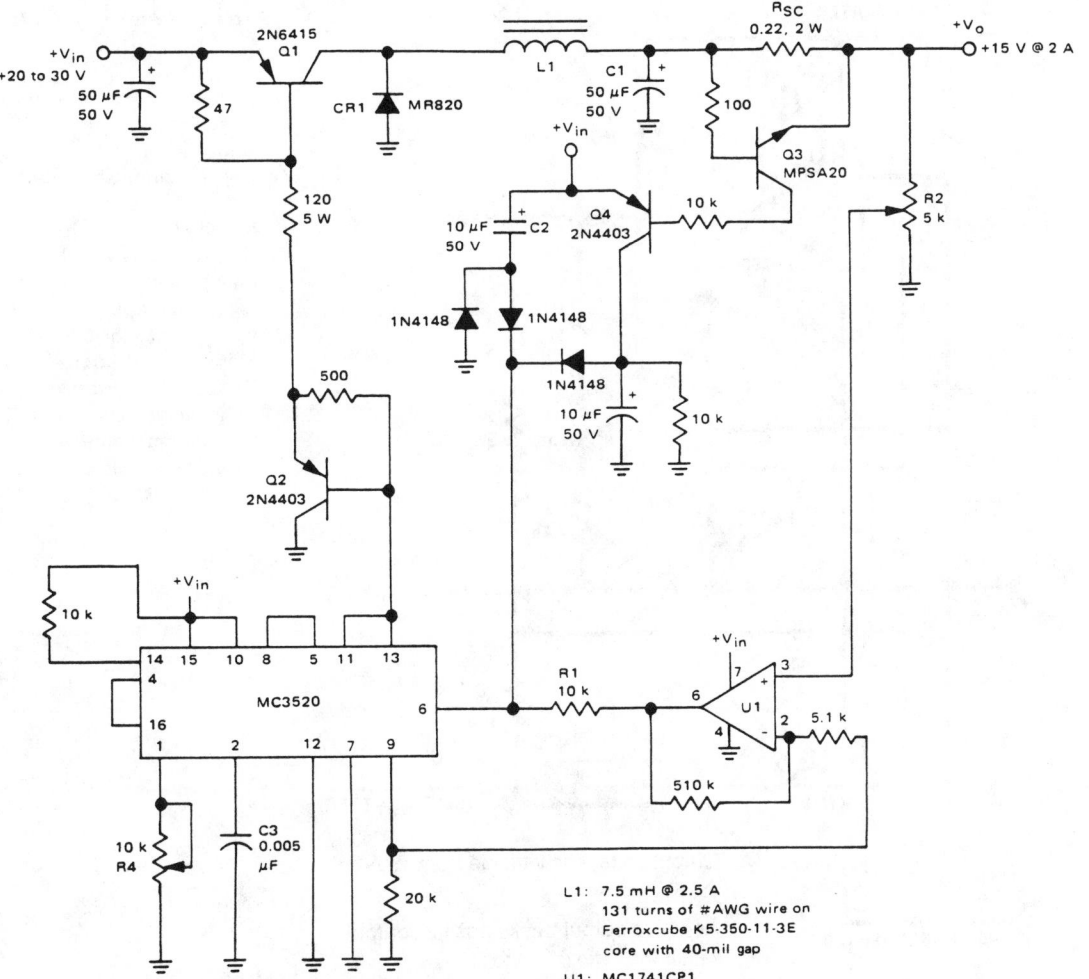

15-Volt, 2-Ampere DC-to-DC Converter. In addition to the op amp, this circuit incorporates the Motorola MC3520 switchmode regulator control IC in a pulse width modulation power supply utilizing a single series switching element. Q1 is the series-switching transistor which chops the DC input at a 25 kHz rate. The resulting waveform is filtered by C1 and L1 to provide a pure DC output. Frequency is fixed via R4 and C3. The voltage output is regulated by comparing its value with that of the reference voltage of the 3520. The op amp is used to amplify the error voltage, and its output is fed into the 3520 to provide pulse width modulation to Q1, thereby controlling the value of the output voltage.—"Motorola Linear/Switchmode Voltage Regulator Handbook." Courtesy of Motorola Semiconductor Products, Motorola, Inc.

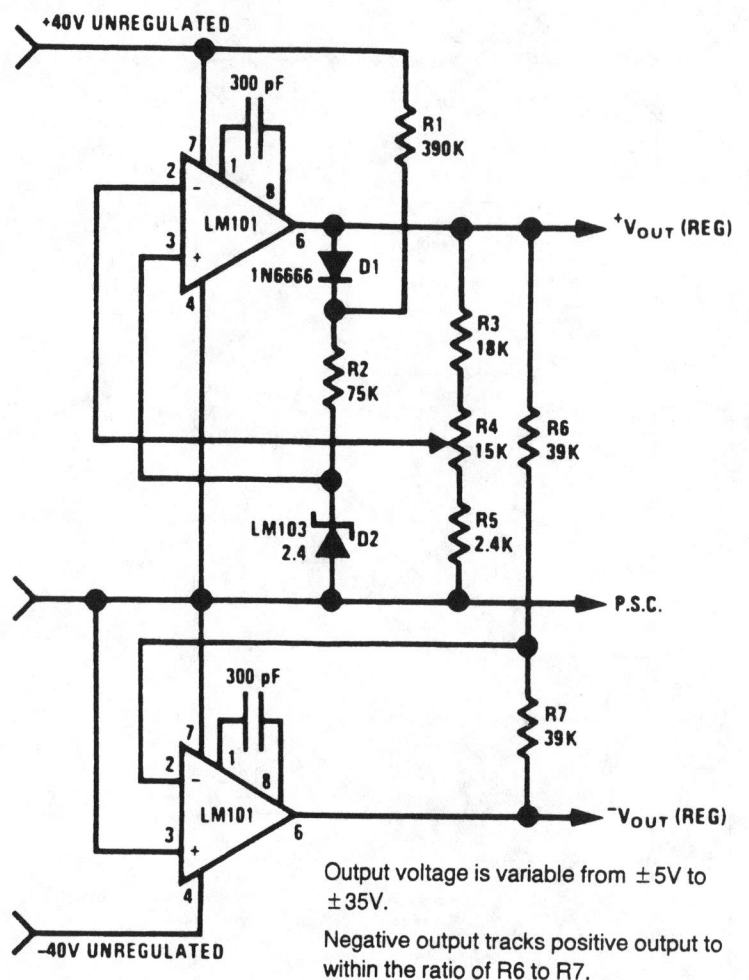

+40V UNREGULATED

300 pF

R1
390K

LM101

1N6666 D1

R3
18K

R2
75K

R4   R6
15K   39K

LM103   D2
2.4

R5
2.4K

+V_OUT (REG)

P.S.C.

300 pF

R7
39K

LM101

-V_OUT (REG)

-40V UNREGULATED

Tracking Power Supply for Op Amps. This power supply circuit overcomes a problem associated with most bipolar power supplies, that of the negative supply potential differing from the potential of the positive supply. With a 40-V DC unregulated bipolar input, regulated output can be varied from 5 to 35 V DC.— "Linear Applications Databook." Courtesy of National Semiconductor Corporation, Santa Clara, California.

Output voltage is variable from ±5V to ±35V.

Negative output tracks positive output to within the ratio of R6 to R7.

Chapter 9

Op Amp Filter Circuits

Several filter circuits utilizing operational amplifiers have already been presented in this text. Filter circuits comprise one of the most basic uses/categories of op amp applications. For the most part, such circuits are quite simple to construct with a single op amp and a handful of discrete RC components. However, these simple circuits are often combined in parallel, cascaded, and generally arranged into complex filter circuits that are capable of a large number of enhancements in a small package.

This chapter is a small part of the total text, as op amp filters are basic circuits that are modified through a change in RC components to address the filtering requirements, roll-offs, passbands, and so on that are desired by the builder. Several circuits are shown that combine simple op amp filters to perform more complex filtering functions.

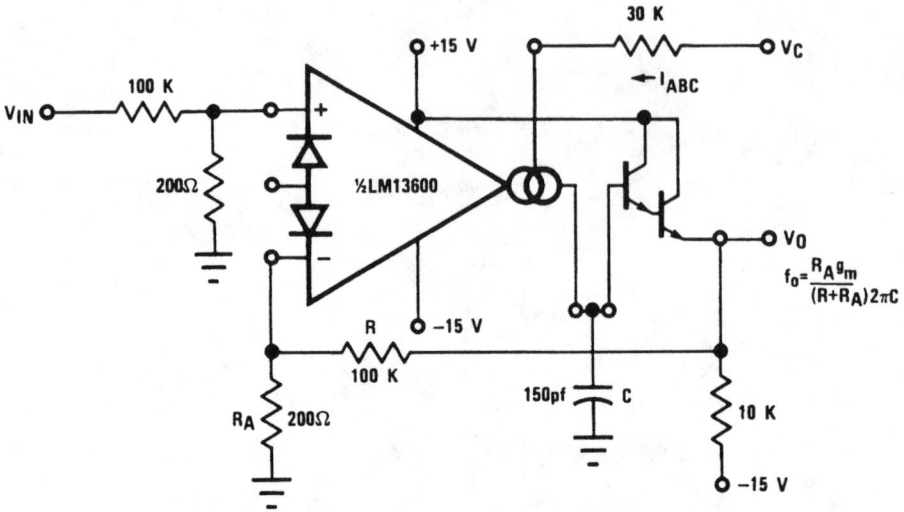

Voltage-Controlled Low-Pass Filter. In this circuit the LM13600 op amp is used as a unity gain or buffer amplifier (amplification factor = 1) at frequencies below the cut-off. At frequencies above cut-off, the circuit offers a single RC roll-off (6 dB per octave) of the amplitude of the input signal. Since only half of the IC is used, with the unused half being identical to this one, this circuit could be easily modified to offer a dual channel low-pass filter, possibly with different cut-offs if desired.—"Linear Databook I." Courtesy of National Semiconductor Corporation, Santa Clara, California.

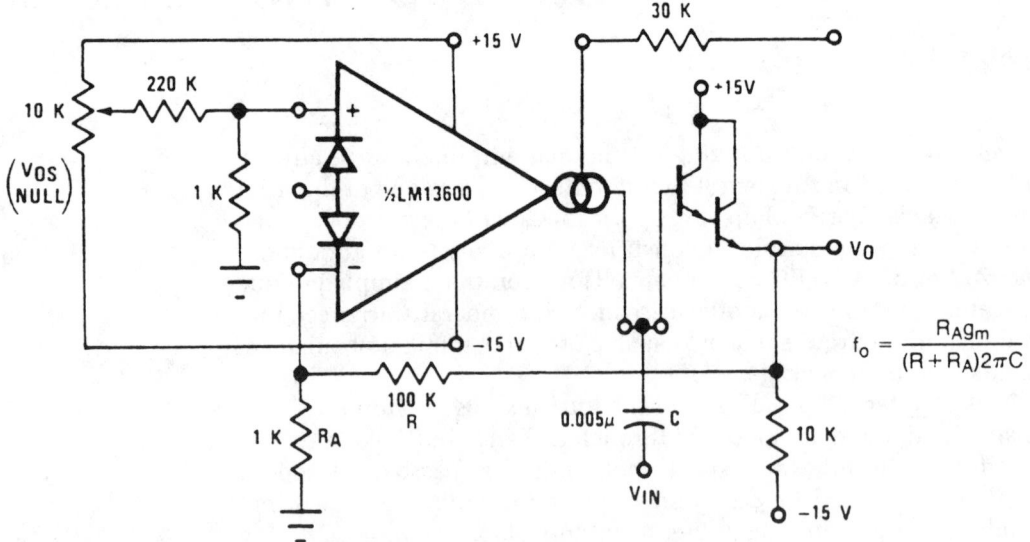

Voltage-Controlled High-Pass Filter. This high-pass filter circuit operates in much the same way as the previous circuit. It provides a single RC roll-off below the cut-off frequency and a unity gain amplifier above the level. The same possibilities for a dual channel version apply.—"Linear Databook I." Courtesy of National Semiconductor Corporation, Santa Clara, California.

Gyrator. This circuit uses the RCA CA3060 tri-operational transconductance amplifier IC connected as a gyrator in an active filter. The amplifiers in this circuit can cause a 3-μF capacitor to function as a floating 10-kH inductor across terminals A and B. A calculated Q for this circuit of 16 will never be achieved, but a Q of 13 is easily attainable in practical circuit constructions. The 20-kΩ to 2-MΩ attenuators in this circuit extend the dynamic range of the amplifier by 100. The 100-kΩ pot across the bipolar supply tunes the inductor by varying the transconductance of the amplifiers, thereby changing the gyration resistance.—"RCA Solid State Databook—Integrated Circuits for Linear Applications." Courtesy of Harris Semiconductor.

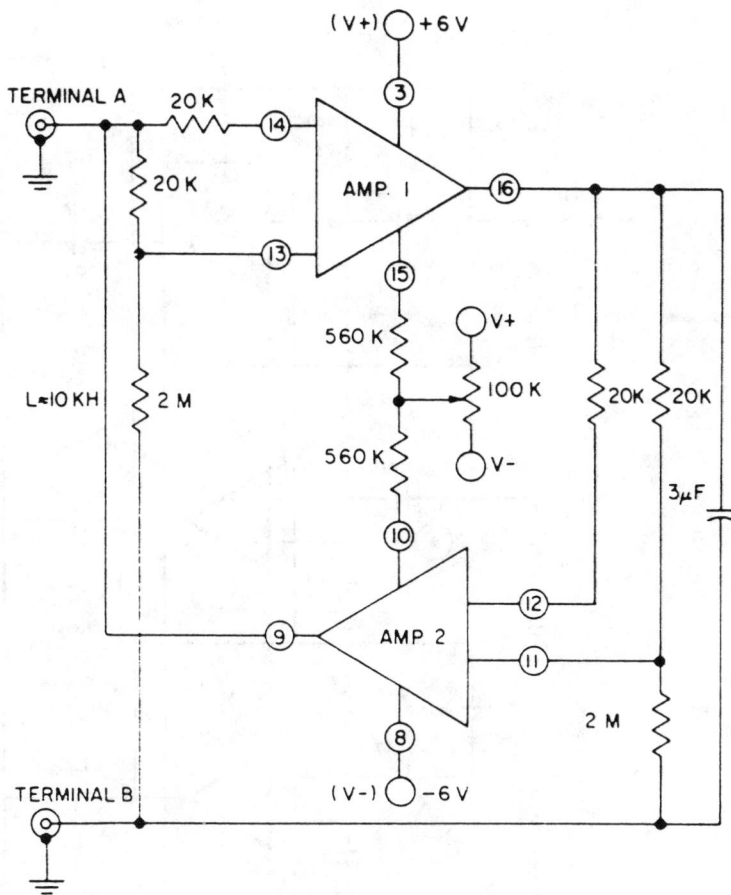

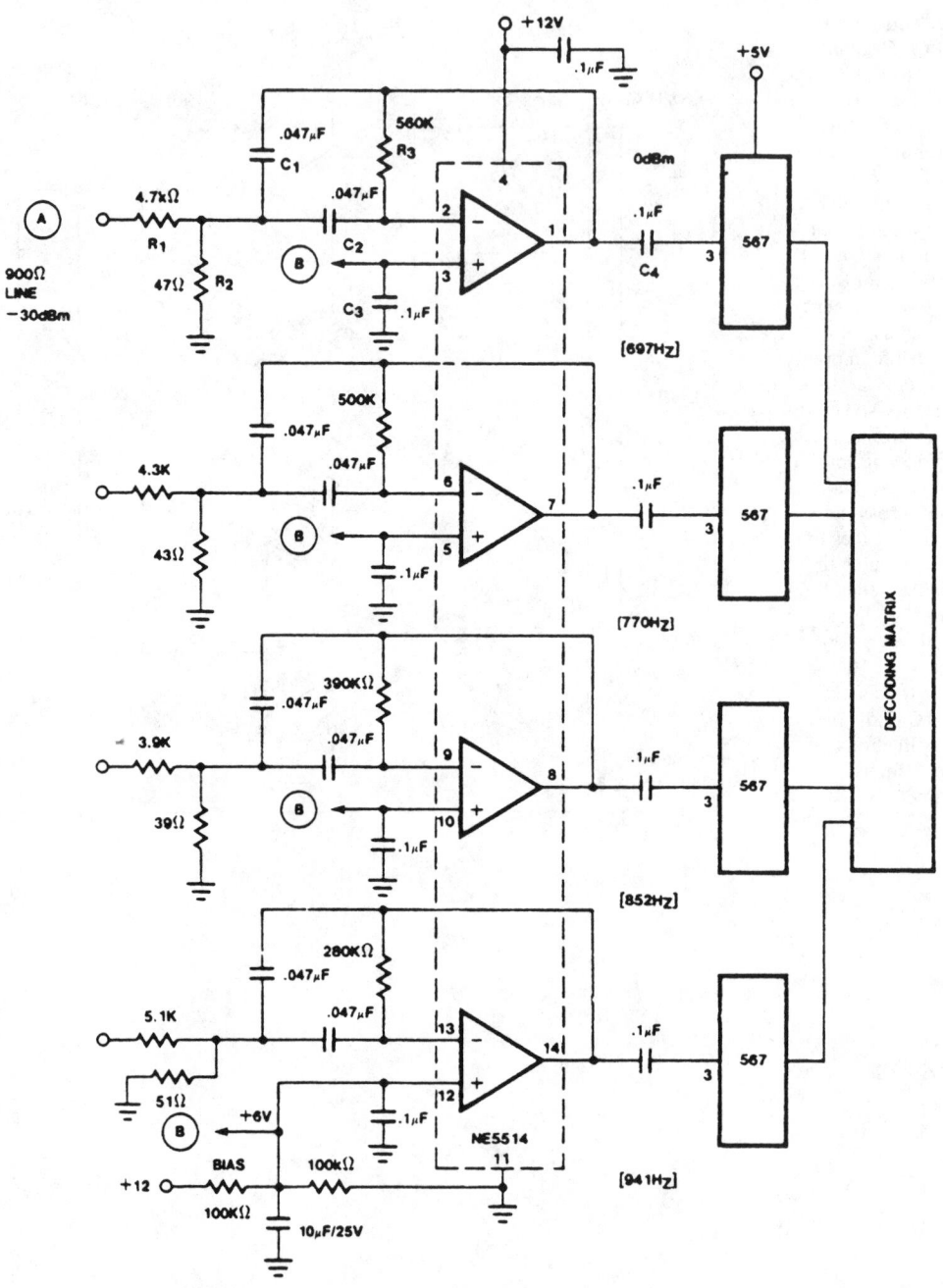

Bandpass Filter for Multichannel Tone Decoder. The use of narrow-band
active prefilters is essential in order to obtain adequate gain and sensitivity in
multiple tone signaling systems that transmit signals over long lines or via
radio carriers. This circuit improves the signal-to-noise ratio and offers four
channels in a single package.—"Linear Data Manual Volume 2: Industrial."
Courtesy of Signetics Company, a division of North American Philips
Corporation.

Chapter 10

Digital Applications

Operational amplifiers are linear integrated circuits, but they are also incorporated in digital circuits, mainly as drivers, selectors, and even references for pure digital devices and/or circuits.

Most of the small amount of circuits presented in this chapter can be described as "committed" to a particular digital device or circuit, although the prolific designer may be able to incorporate these basic designs into other equipment or interface these circuit ideas with other digital devices.

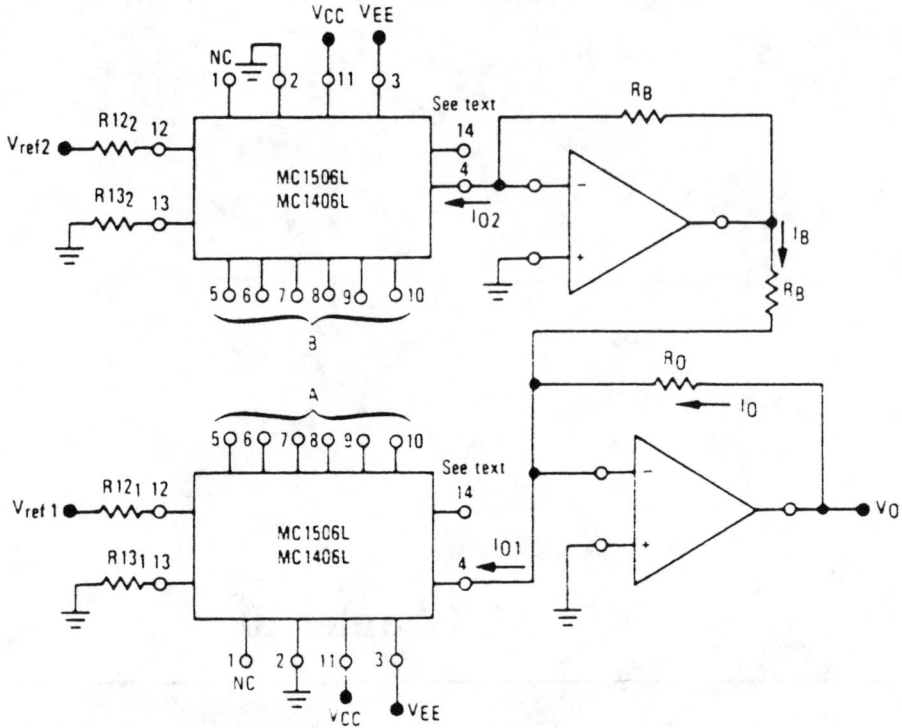

Digital Attenuator. This circuit combines an op amp with the MC1406L/
1506L 6-bit, multiplying ADC from Motorola. This may also be thought of as a
programmable gain amplifier. If $R_s = 50\ \Omega$, then no compensation capacitor is
needed and a wide, large signal bandwidth is achieved.—"Motorola Linear
and Interface Integrated Circuits." Courtesy of Motorola Semiconductor
Products, Motorola, Inc.

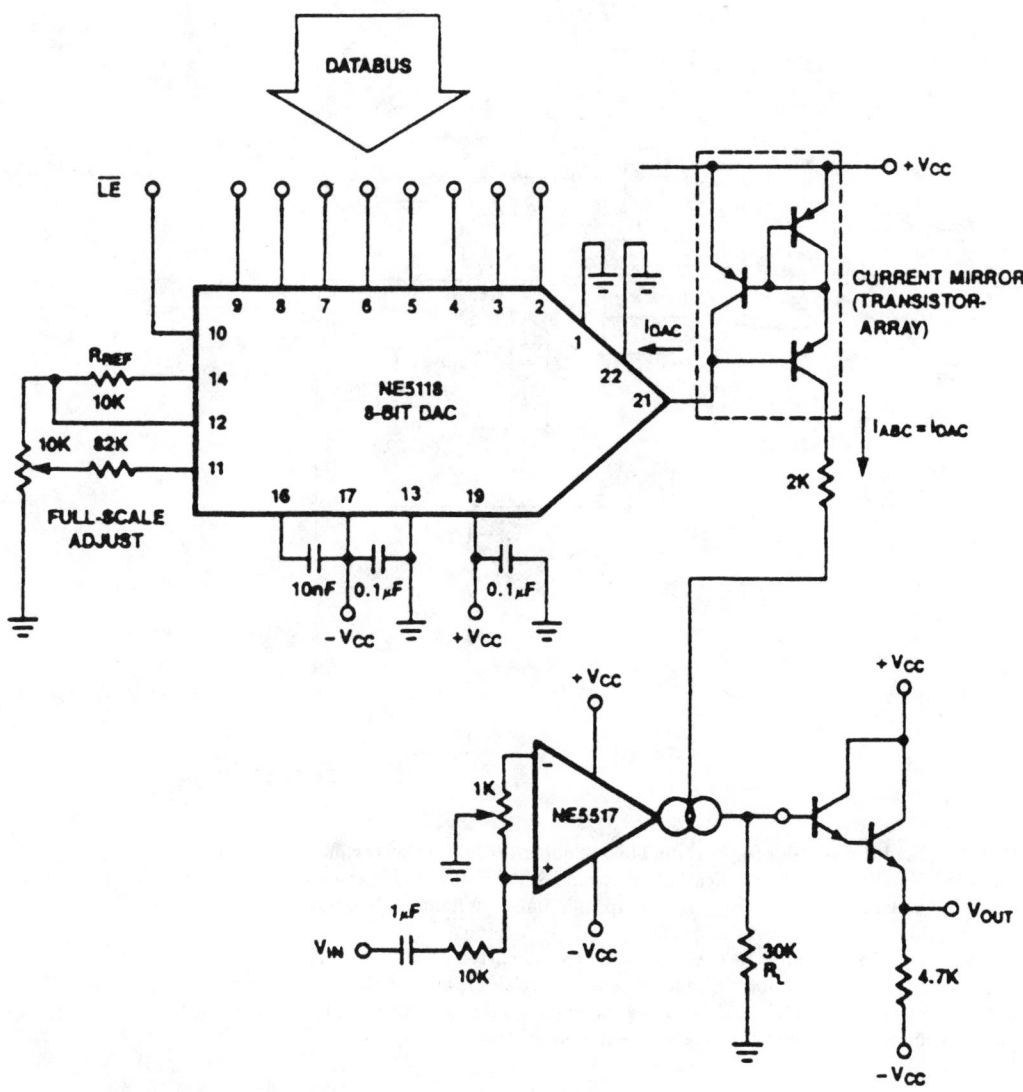

Digitally Programmable Amplifier. This circuit consists of a Signetics microprocessor-compatible digital-to-analog converter, a transistor array, and the NE5517 operational transconductance amplifier (configured as a voltage-controlled amplifier).—"Linear Data Manual Volume 2: Industrial." Courtesy of Signetics Company, a division of North American Philips Corporation.

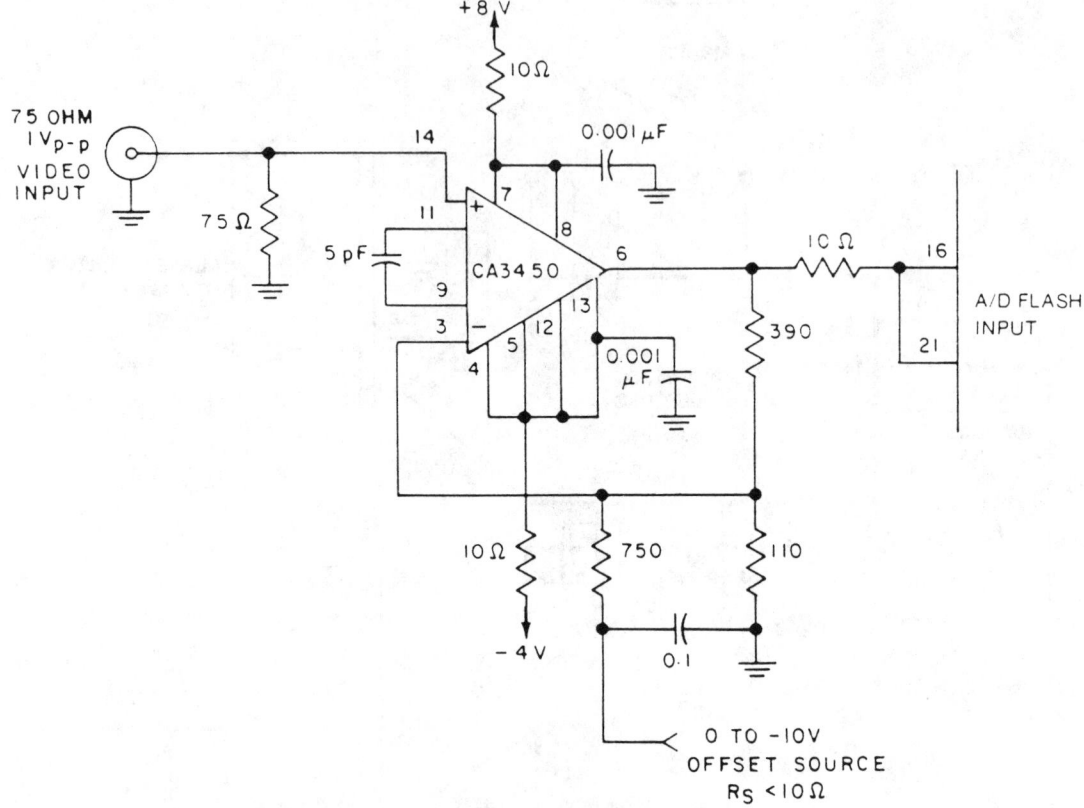

High-Bandwidth Amplifier for Driving Flash Analog-to-Digital Converter.
This circuit was designed specifically to drive the RCA CA3318 Flash analog-
to-digital converter, a device designed for applications that demand low power
consumption and high-speed digitization. A 1-V peak-to-peak signal is
required at the 75-Ω input. The CA3450 is a video line driver, high-speed op
amp exhibiting high open-loop gain at frequencies to 5 MHz and a power
bandwidth of 10 MHz.—"RCA Solid State Databook—Integrated Circuits for
Linear Applications." Courtesy of Harris Semiconductor.

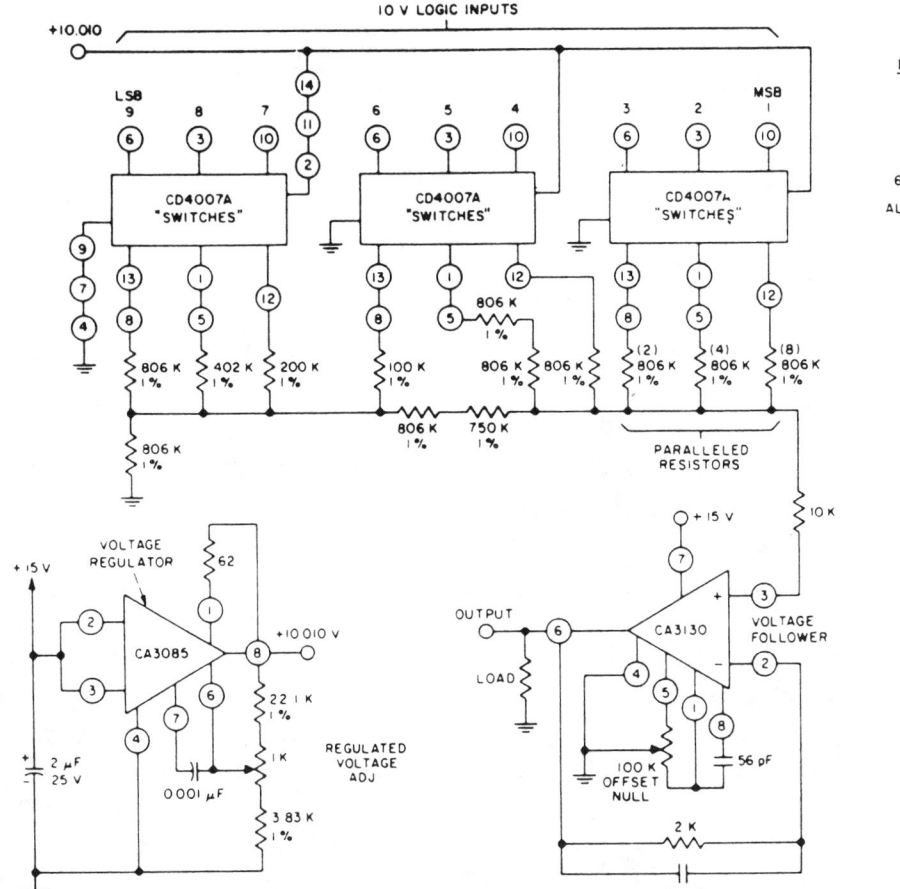

9-Bit COS/MOS Digital-to-Analog Converter. This circuit combines the concepts of multiple-switch CMOS ICs a low-cost ladder network of discrete metal-oxide-film resistors, an RCA CA 3130 op amp connected as a follower and an inexpensive monolithic regulator in a single power supply arrangement. The single 15-V DC supply provides a positive bus for the follower amplifier and also feeds the voltage regulator. The output from this regulator (10 V) provides a scale-adjust function.—"RCA Solid State Databook—Integrated Circuits for Linear Applications." Courtesy of Harris Semiconductor.

Chapter 11

Miscellaneous Circuits

This chapter of the text is devoted entirely to those operational amplifier circuits that, for one reason or another, did not quite fit into any of the previous chapters. This then is the proverbial "melange" of op amp circuits that address a variety of needs and applications, although some may have uses that are very obscure (or arguably non-existent).

Here, the reader will find AM modulators, double-sideband generators/detectors, biasing circuit schemes, a Kelvin sensing circuit, and a micropower op amp booster. Of course, the hands-down favorite has to be a circuit named "Son of Godzilla Op Amp Booster." This is an interesting circuit with truly *lethal* potentials and, admittedly, one that has little practical use outside of industrial fields. Nevertheless, this and many other circuits that span the spectrum from obscure to useful to fanciful are covered in this chapter.

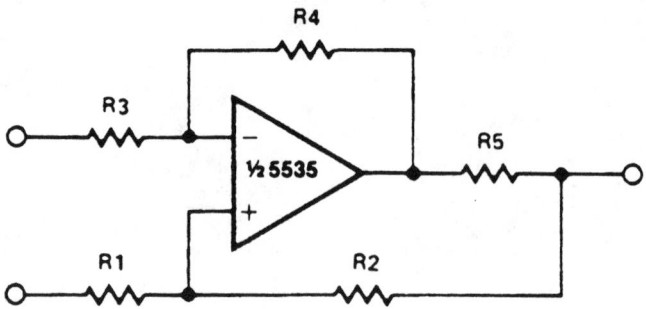

NOTES:

$$\frac{R_2}{R_1} = \frac{R_4}{R_3}$$

$$I_{OUT} = \frac{V_{IN} \cdot R_2}{R_5 \cdot R_1}$$

Voltage-to-Current Converter. This circuit has two inputs and will produce either polarity of output current. However, a severe disadvantage is present in the form of the error current that flows through R2 and the current limit available. There are better ways to accomplish this conversion without the same limitations.—"Linear Data Manual Volume 2: Industrial." Courtesy of Signetics Company, a division of North American Philips Corporation.

Differential to Single-Ended Converter.   This circuit converts differential inputs into single ended outputs using a minimum of external components. The RCA CA3280 offers an excellent common-mode rejection ratio and allows matching resistor networks to be avoided. This circuit handles input signals in the ±25-mV range.—RCA Solid State Databook—Integrated Circuits for Linear Applications." Courtesy of Harris Semiconductor.

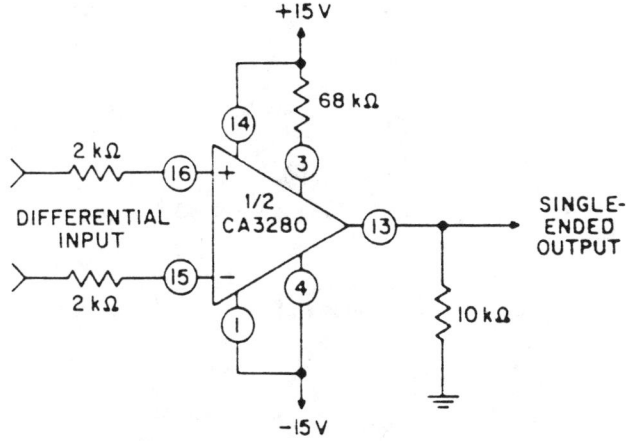

Magnetic Transducer Detector. This is a zero-crossing detector for a magnetic transducer such as a magnetometer or shaft-position pick-off. It delivers the output signal directly to TTL or DTL logic circuits. The power supply is a unipolar type and it is assumed that the circuit will be powered by the logic supply. R1 and R2 form a resistive divider which biases the inputs 0.5 V above ground. An optional offset balancing circuit is formed by R3 and R4.—"Linear Applications Databook." Courtesy of National Semiconductor Corporation, Santa Clara, California.

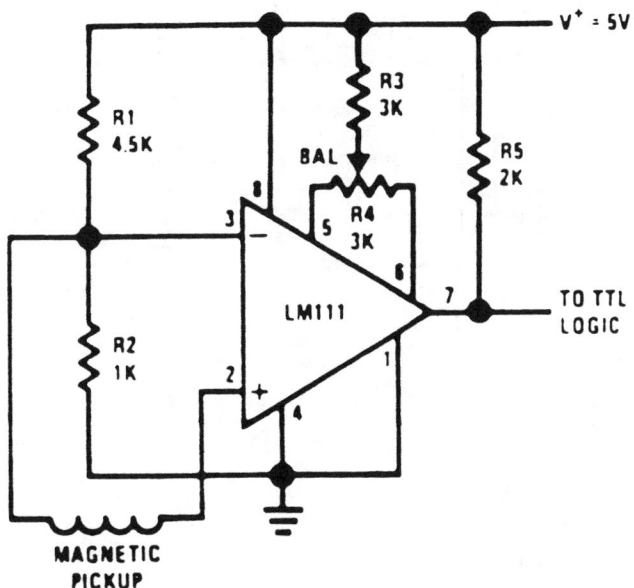

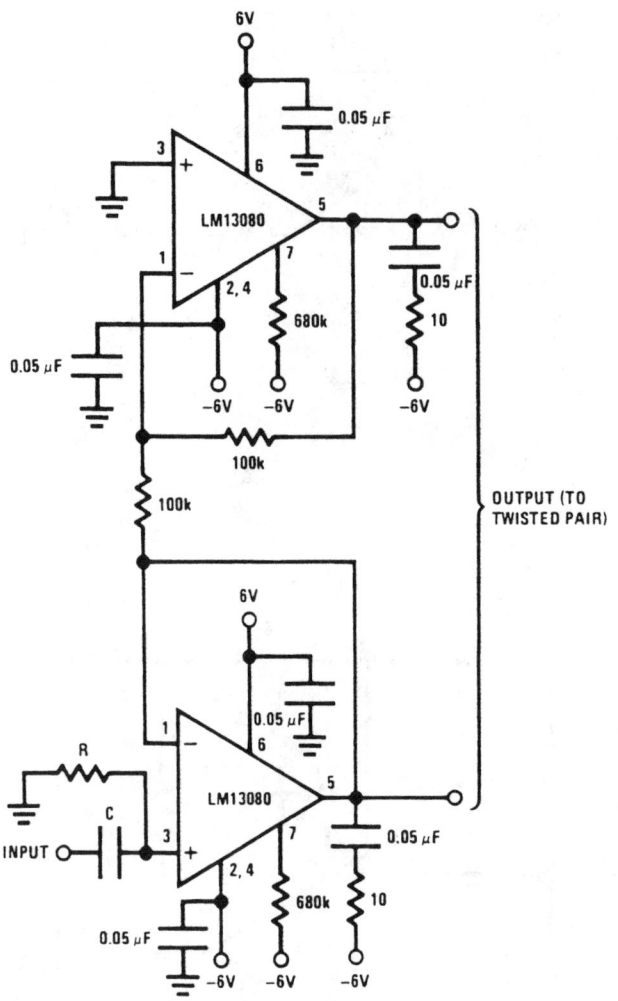

Line Driver. This circuit produces a balanced output suitable for driving low-impedance lines from an unbalanced, high-impedance input. This is accomplished by connecting two LM13080 programmable power op amps in an opposite polarity output configuration. The bandwidth is good, and this circuit will deliver a 20-V peak-to-peak output into a 50-Ω load at frequencies up to 10 kHz and it will still provide as much as a 13-V peak-to-peak signal at 20 kHz.—"Linear Databook I." Courtesy of National Semiconductor Corporation, Santa Clara, California.

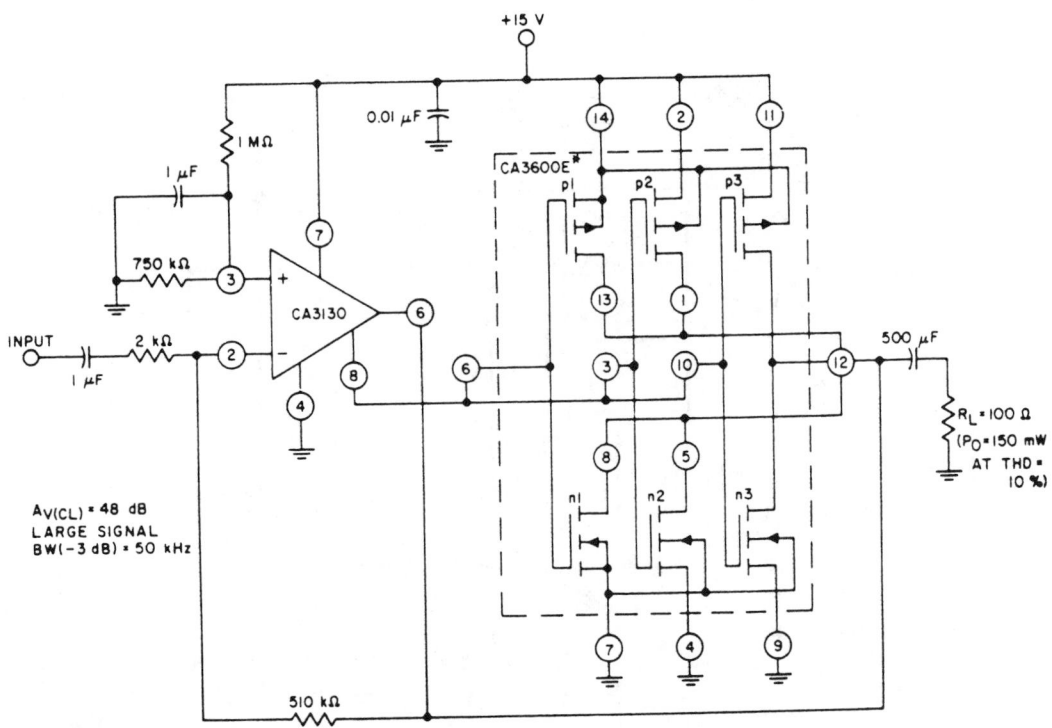

Power Booster for Op Amps.   Here an RCA CA3600E CMOS array is connected to the output stage of the op amp to produce an output of 150 mW into a 100-Ω load. This circuit uses a total of ten, discrete components and can be assembled on a piece of vector board in a very short time. Typically, this circuit will consume about 20 mA at a supply potential of 15 V DC. It boosts the current-handling capability of the op amp output stage by a factor of 2.5.— "RCA Solid State Databook—Integrated Circuits for Linear Applications." Courtesy of Harris Semiconductor.

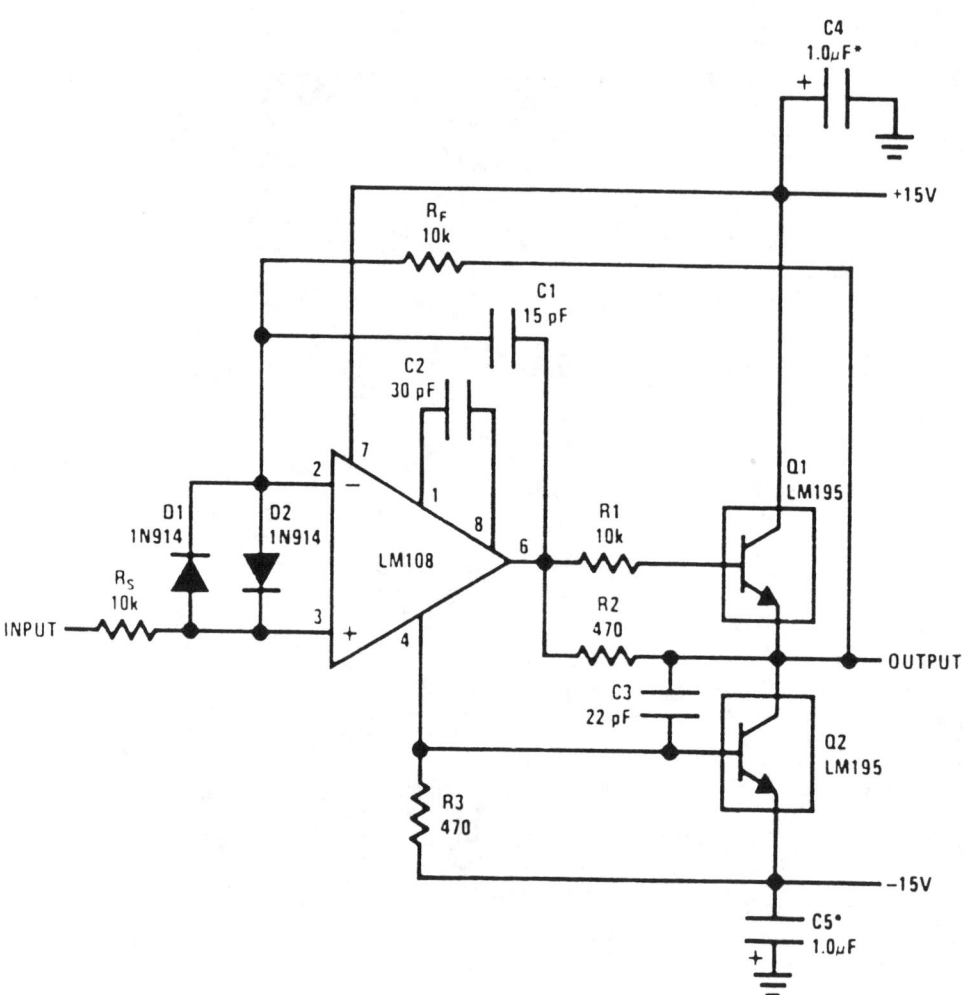

1.0-Ampere Voltage Follower.   The power of the standard voltage follower configuration of this op amp is enhanced by the addition of Q1 and Q2 which are National Semiconductor LM195 monolithic power transistor packages. C5 should be a solid tantalum type of device.—"Linear Databook III." Courtesy of National Semiconductor Corporation, Santa Clara, California.

Universal Offset Null for Inverting Amplifiers. A large number of op amps contain the necessary pin connections to provide external offset adjustments. Many others do not. This circuit allows for such adjustments in the latter type of inverting amplifer.— "Linear Data Manual Volume 2: Industrial." Courtesy of Signetics Company, a division of North American Philips Corporation.

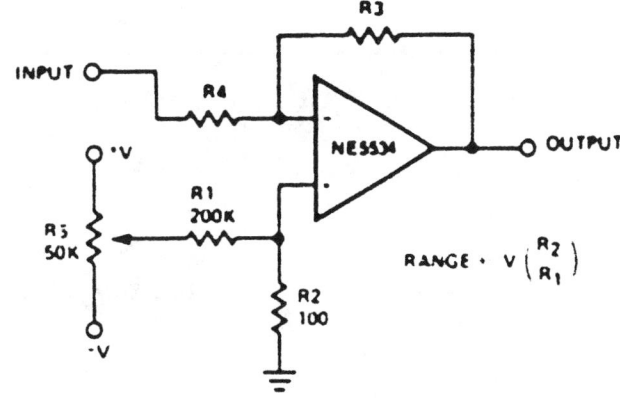

Universal Offset for Noninverting Amplifiers. This is the noninverting equivalent of the previous circuit.—"Linear Data Manual Volume 2: Industrial." Courtesy of Signetics Company, a division of North American Philips Corporation.

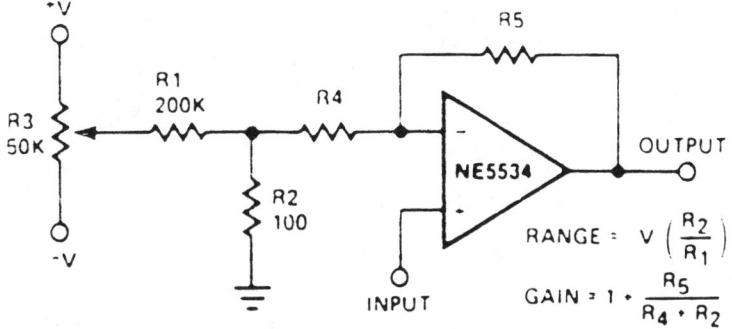

Bias Current Compensation. This circuit provides sufficient current into the input to cancel the bias current requirement.— "Linear Data Manual Volume 2: Industrial." Courtesy of Signetics Company, a division of North American Philips Corporation.

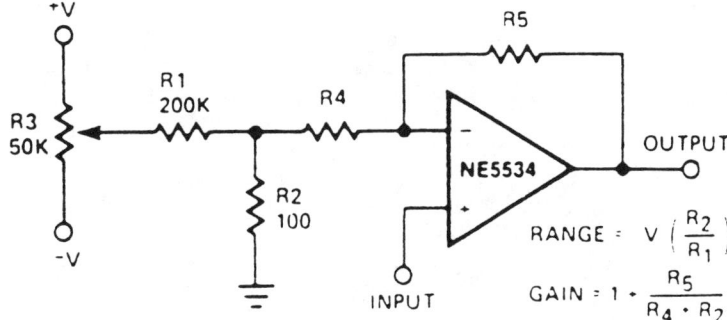

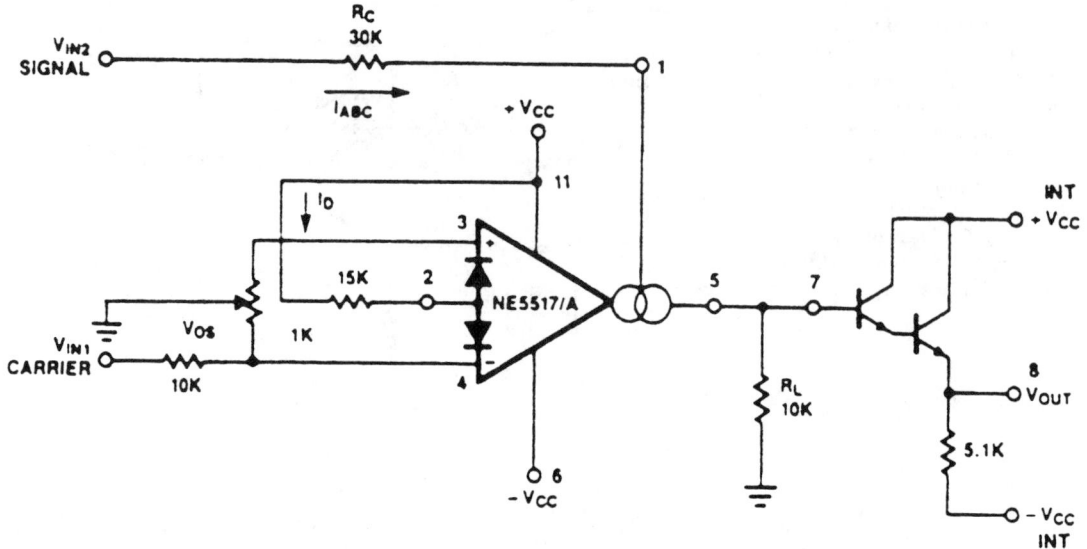

Amplitude Modulator. The NE5517 is an operational transconductance amplifier meaning that the transconductance of this device is directly proportional to the current. Due to this characteristic, the amplification of a signal can be easily controlled. The output current is the product of input voltage times transconductance. This circuit should be practical up to approximately 225 kHz with 99% modulation easily achieved.—"Linear Data Manual Volume 2: Industrial." Courtesy of Signetics Company, a division of North American Philips Corporation.

Amplitude Modulator. This amplitude modulator is little more than one-half ▶ of a stereo volume control discussed elsewhere in this text. It uses one-half of the LM13700 dual transconductance op amp. The carrier signal is applied to the inverting input while the modulation information is applied at the output. Again, you can simply drop the lower half of the stereo volume control circuit using this same IC and end up with the amplitude modulator. There are no changes in external component values.—"Linear Databook I." Courtesy of National Semiconductor Corporation, Santa Clara, California.

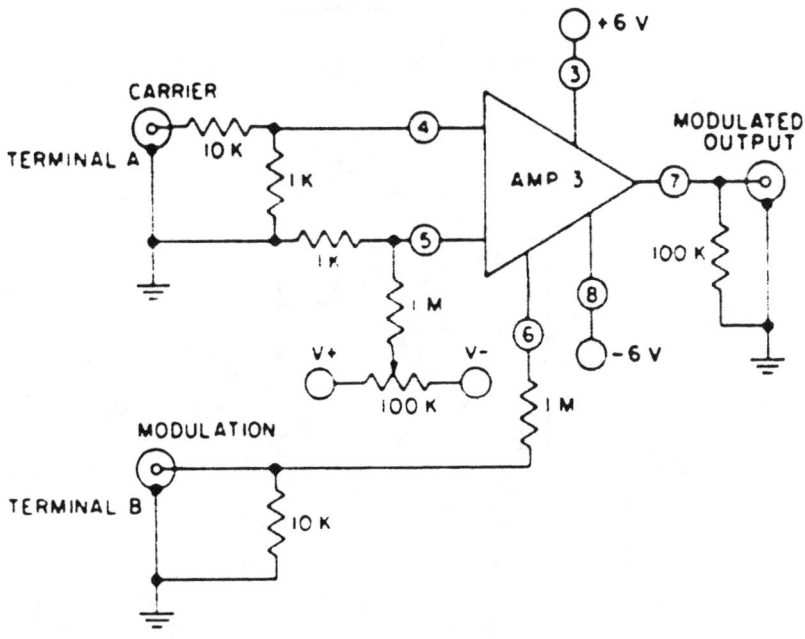

AM Modulator. This circuit uses one of three op amps available in the RCA 3060 IC as an AM modulator. Modulation is applied to the amplifier bias input (terminal B), and the carrier frequency is applied to the differential input at terminal A.—"RCA Solid State Databook—Integrated Circuits for Linear Applications." Courtesy of Harris Semiconductor.

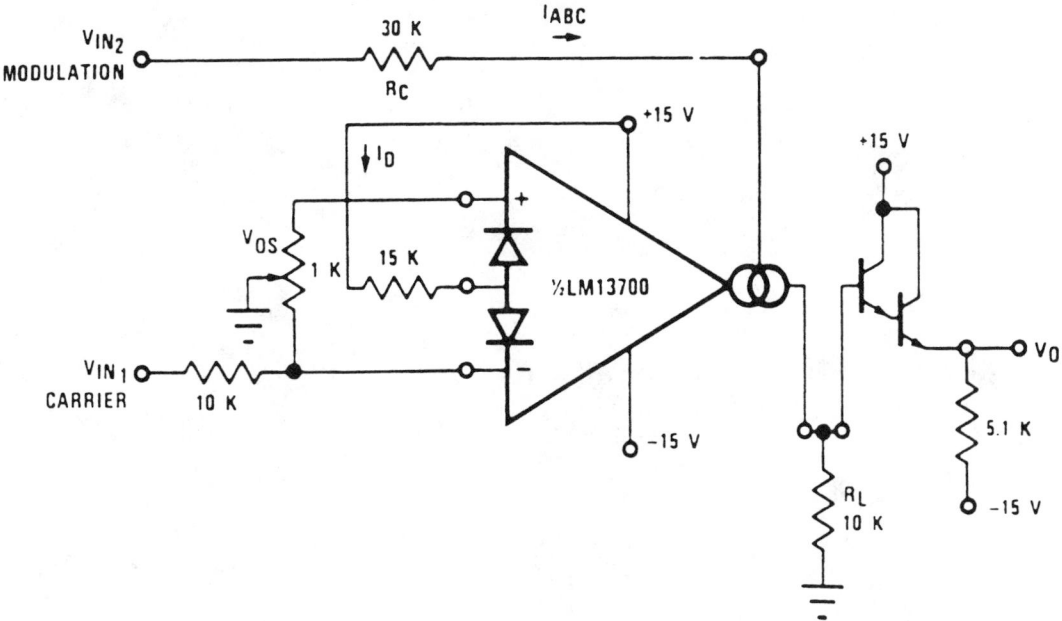

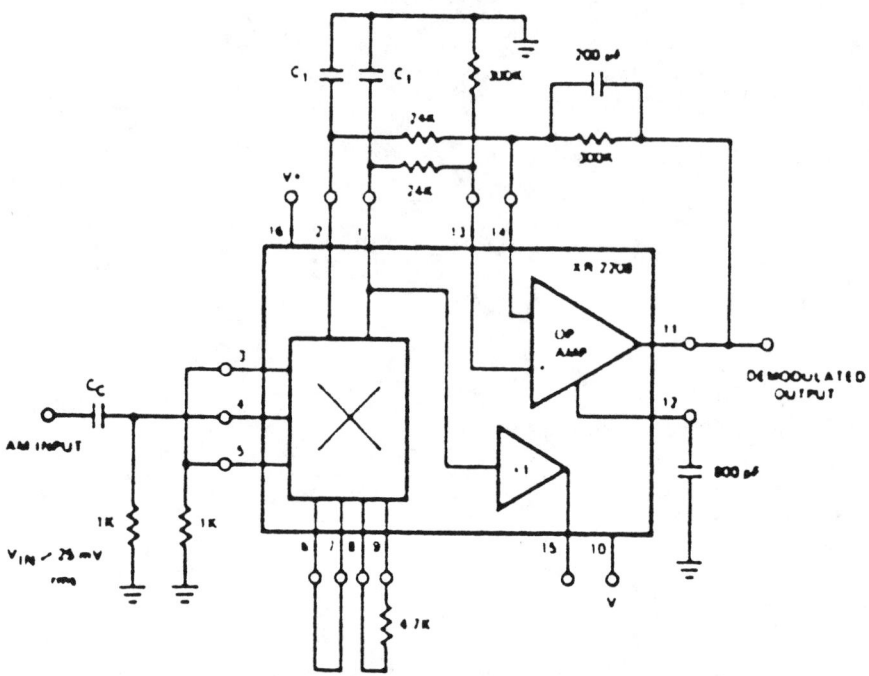

AM Detector. This synchronous AM detector operates at up to 100 MHz. The input signal is applied to pin 4. With the Y-gain terminals shorted, the multiplier section of this IC serves as a limiter. Low-pass capacitors (C1) filter the carrier feedthrough. The op amp section can be utilized as an audio preamplifier to increase the demodulated output amplitude.—"EXAR Databook." Courtesy of EXAR Corporation.

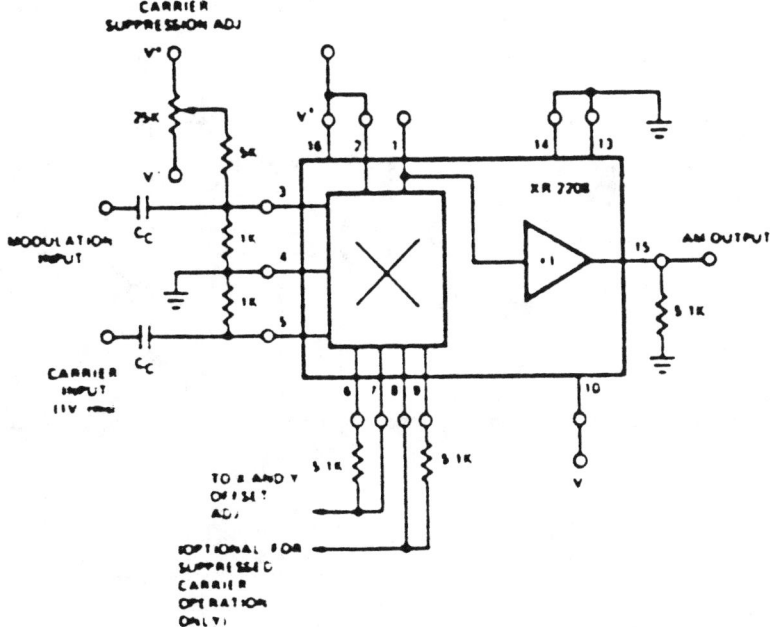

AM or DSB Generator. This circuit uses the EXAR operational multiplier which combines a 4-quadrant analog multiplier, a high-frequency buffer amplifier, and an op amp on the same chip. Modulation and carrier inputs are applied to the X and Y inputs, respectively. Carrier level output is adjusted by the DC voltage at pin 3. For suppressed carrier operation, the carrier feedthrough is reduced by using the X and Y offset adjustments. Typical suppression without offset is 40 dB at 1 MHz and 30 dB at 10 MHz.—"EXAR Databook." Courtesy of EXAR Corporation.

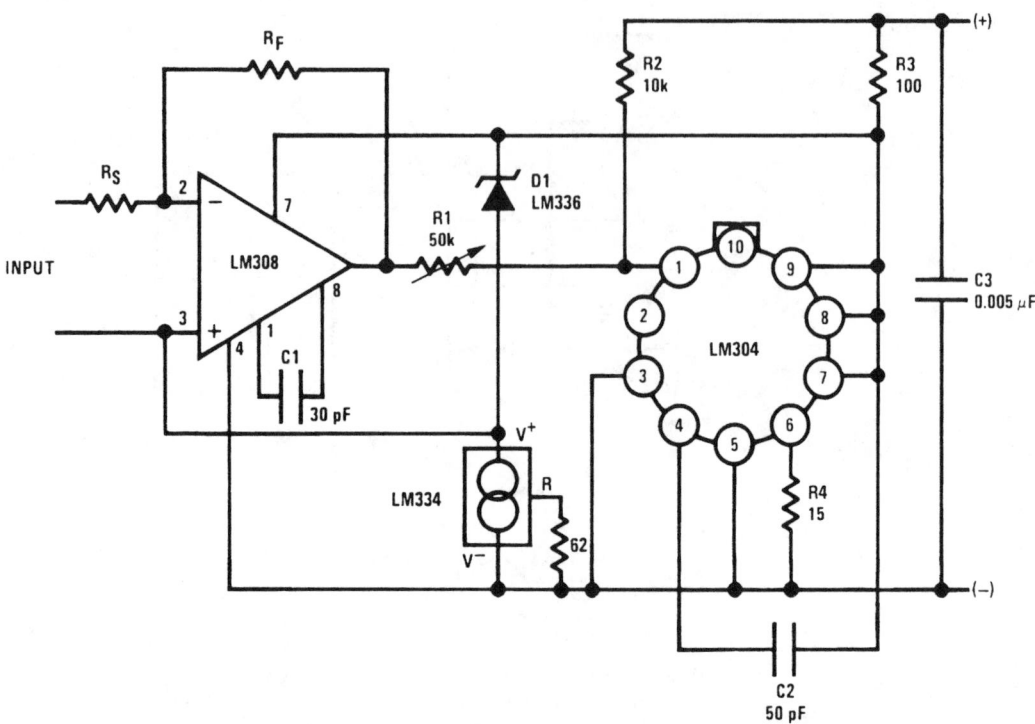

Current Transmitter and Amplifier. A standard current transmitter configuration using the LM304 can be coupled to an LM308 op amp for additional gain as this circuit demonstrates. The output of the op amp drives the LM304 through R1 which is used to set the zero input current in conjunction with D1, the reference diode. Linearity is better than 0.1%. Current change with voltage is approximately 7 μA/V.—"Linear Applications Databook." Courtesy of National Semiconductor Corporation, Santa Clara, California.

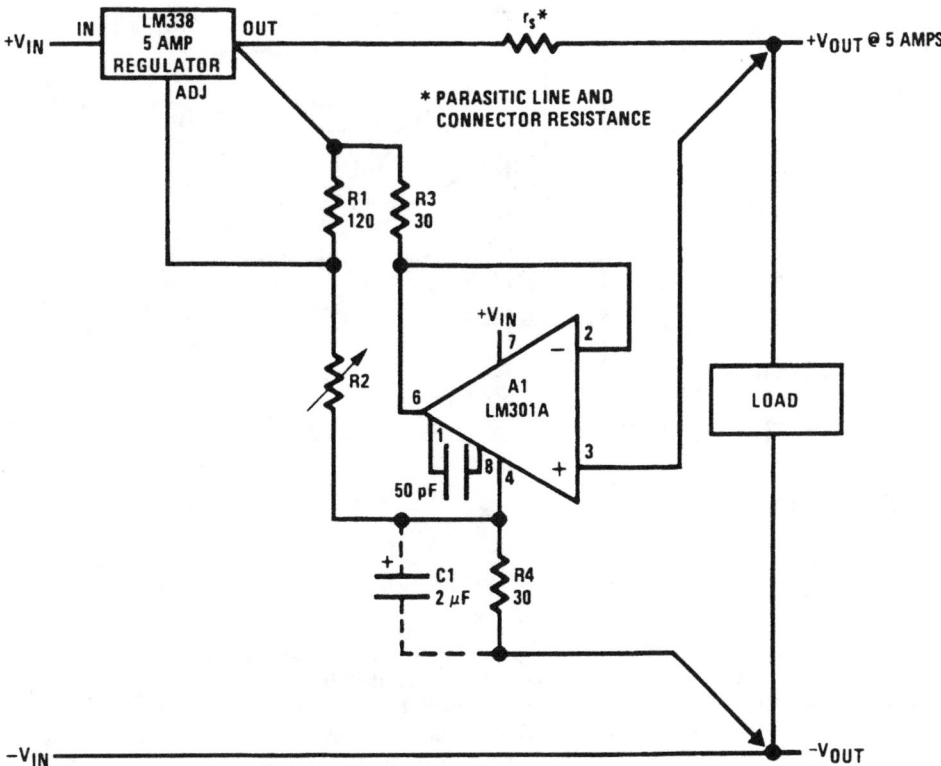

Kelvin Sensing Circuit. A new technique for current sharing connections of three terminal regulators is demonstrated by this Kelvin sensing circuit. A1 forces a voltage drop across R3 equal to the voltage across the parasitic resistance R_s. Current through R3 flows into the output of A1 and out the negative supply pin, creating a voltage drop across R4 that is just equal to the voltage across R_s. This cancels the effect of R_s on load regulation. This is a far better method than inserting low value resistors in series with the output lines from each of the parallel regulators. C1 is necessary only if intermediate values of capacitance (2 through 20 μF) are placed across the load. R_s1 through R_s3 are equal lengths of #22 gauge lead wire which act as ballasting resistors.— "Linear Applications Databook." Courtesy of National Semiconductor Corporation, Santa Clara, California.

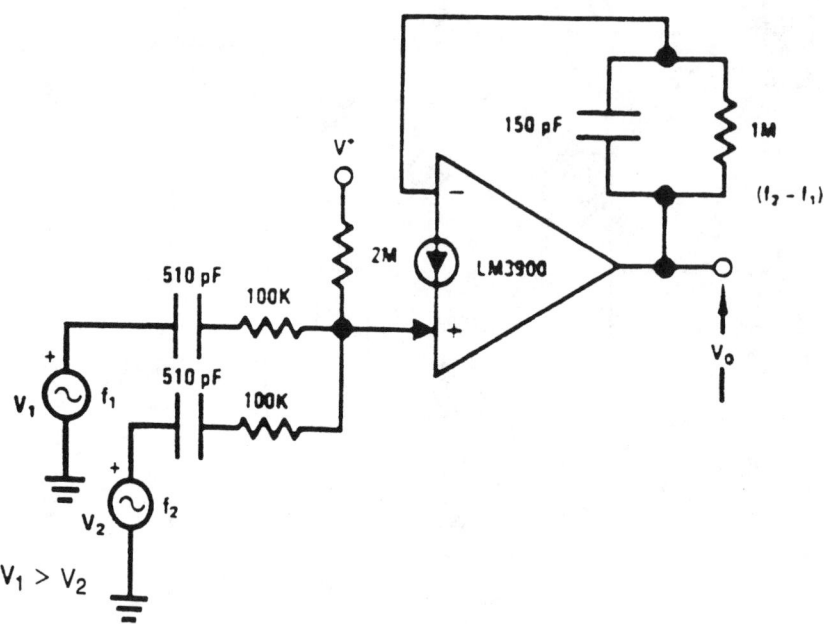

Low-Frequency Mixer.   The LM3900 quad op amp can be used for nonlinear
signal processing due to the fact that a diode exists at its positive input
terminal. This simple, low-frequency mixer is a result of this construction and
may be used at relatively high frequencies as long as the difference in the two
frequencies is within the bandwidth capabilities of the op amp. This mixer has
a gain of 10 and a corner frequency of 1 kHz. The local oscillator is determined
by the input source with the highest amplitude. When a smaller signal is
added at the second input, the difference frequency is filtered from the
complex waveform and is made available at the output.—"Linear Applications
Databook." Courtesy of National Semiconductor Corporation, Santa Clara,
California.

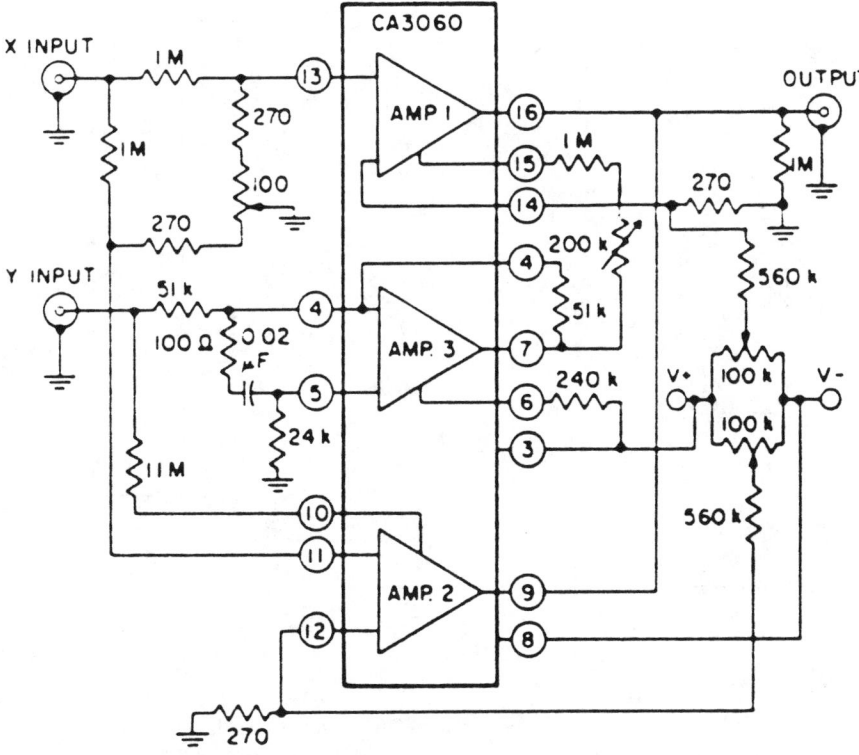

Four-Quadrant Multiplier. The three operational transconductance amplifiers in the RCA 3060 IC are put to good practical use in this circuit. Amplifier 1 is connected as an inverting amplifier for the X-input signal. Amplifier two is noninverting, while amplifier three is connected as a unity gain inverting amplifer.—"RCA Solid State Databook—Integrated Circuits for Linear Applications." Courtesy of Harris Semiconductor.

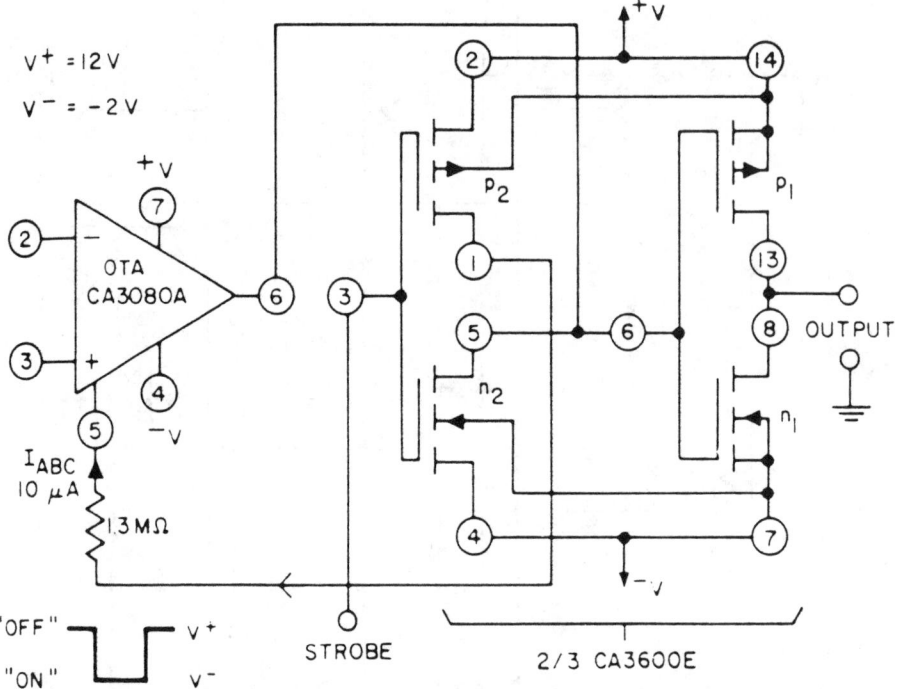

Programmable Micropower Comparator.   An RCA CA3600E COS/MOS array
is used in this circuit which requires about 10 μW of power in its resting state.
When the comparator is strobed *on*, P1 is driven into conduction and the
operational transconductance amplifier becomes active, consuming
approximately 400 μW. Voltage gain is typically 130 dB.—"RCA Solid State
Databook—Integrated Circuits for Linear Applications." Courtesy of Harris
Semiconductor.

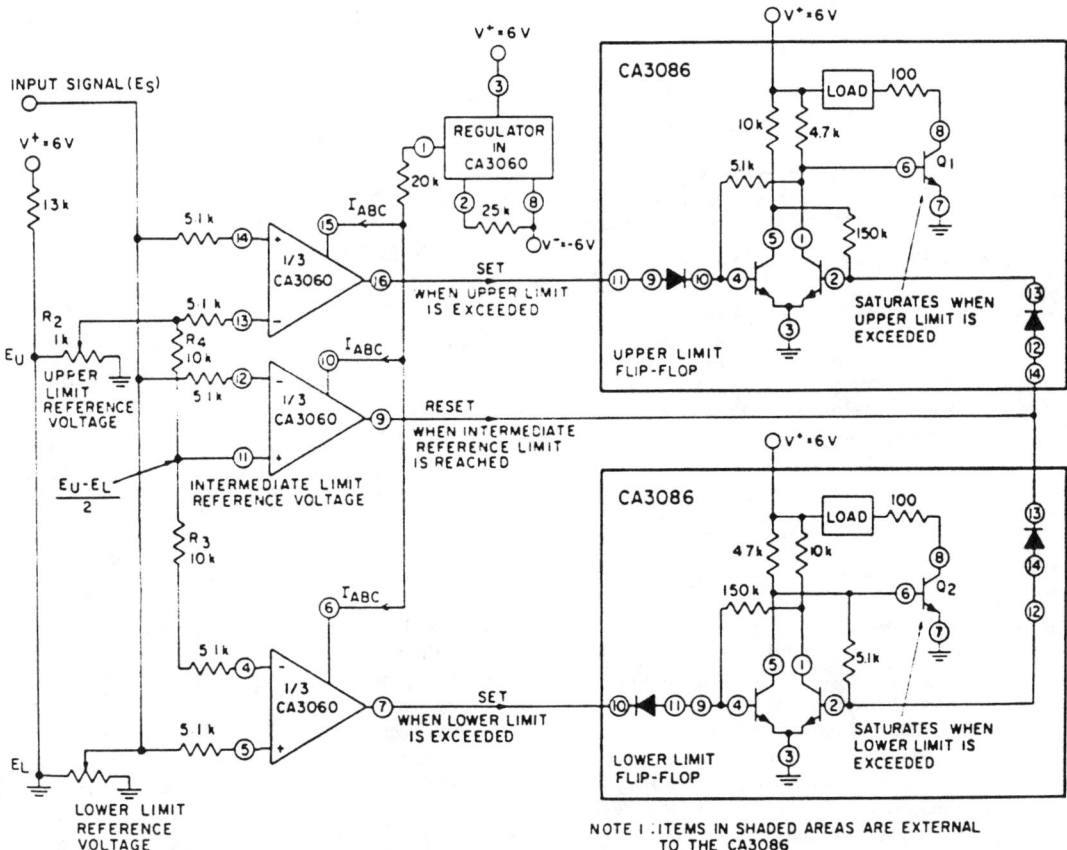

Tri-Level Comparator Circuit. A tri-level comparator circuit is an ideal application for a multioperational amplifier package that contains three op amps on a single chip. Such is the case with the RCA 3060. In a three level comparator, there are three adjustable limits. If either the upper or the lower limit is exceeded, the appropriate output is activated until the input signal is returned to a preselected intermediate value. The built-in regulator of this chip provides bias current to the three amplifiers from the bipolar 6-V DC supply. The lower- and upper-limit reference voltages are selected by adjusting R1 and R2. When R3 and R4 are of the same value, the intermediate-limit reference voltage is automatically established at a mean value between these two limits. Other values may be selected by varying the values of these two resistors. In this circuit, 5-V, 25-mA lamps serve as loads.—"RCA Solid State Databook— Integrated Circuits for Linear Applications." Courtesy of Harris Semiconductor.

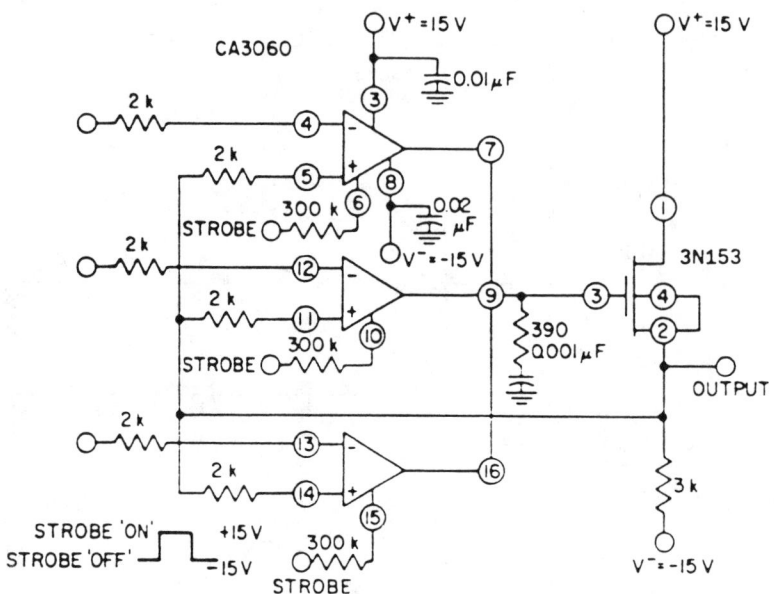

Three-Channel Multiplexer. This three-channel multiplexer uses the tri-amp arrangement in the RCA 3060 operational transconductance amplifier IC. The circuit uses the CA3060 amplifiers in a high-input impedance voltage follower and in a cascade arrangement. This provides an open-loop voltage gain in excess of 100 dB and assures excellent accuracy in the voltage follower mode with 100% feedback. This circuit operates from a dual polarity 15-V DC supply but may be operated from 6 V DC by altering the bias current series resistance and decreasing the drain resistance for the MOS/FET. This circuit uses a phase compensation network comprised of the 390-Ω resistor and a 1000-pF capacitor. System bandwidth is 1.5 MHz.—"RCA Solid State Databook—Integrated Circuits for Linear Applications." Courtesy of Harris Semiconductor.

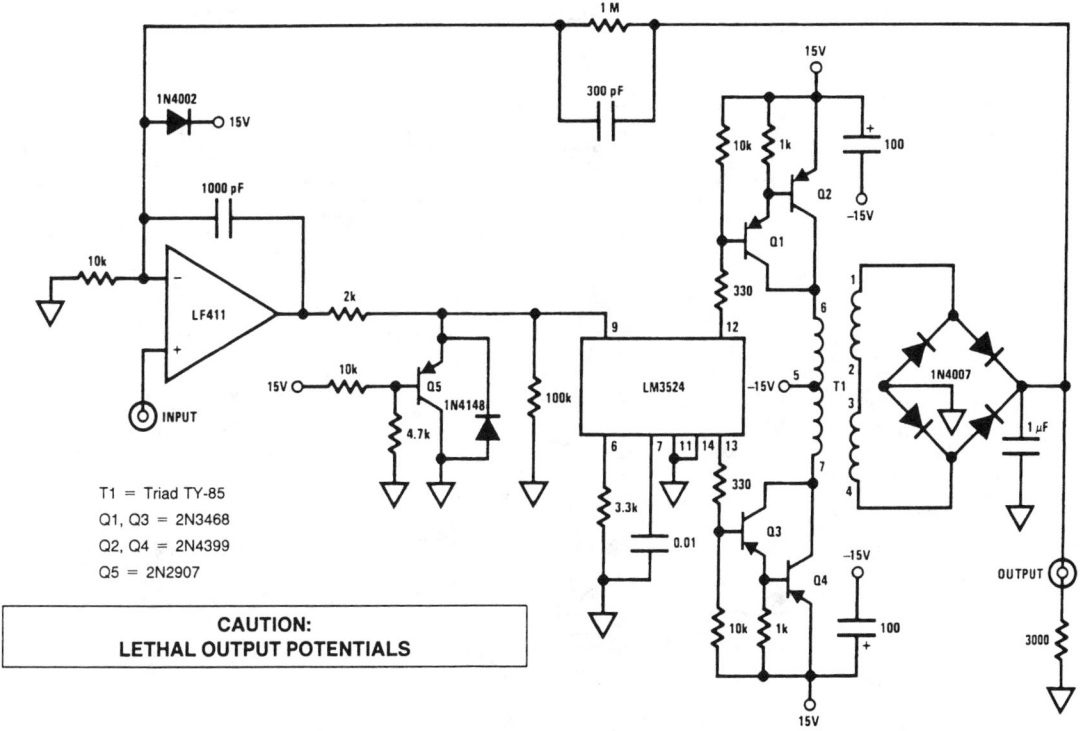

Son of Godzilla Op Amp Booster. The author doubts the practicality of this circuit from National Semiconductor, but the name makes it mandatory for inclusion in this text. Put simply, this is a very-high-voltage, high-current booster circuit that will allow an op amp to control up to 300 W for positive output potentials of up to 1000 V. This booster operates in a switching mode, so it exhibits very high efficiency. It operates from a bipolar 15-V DC supply, so it does not require a high-voltage power supply to achieve its high-voltage potential. The LM3524 switching regulator chip is used to pulse width modulate the transistors. They, in turn, provide a 20-kHz drive to T1, a step-up transformer. The rectified and filtered output from T1 is fed back to the LF411 op amp which controls the input to the LM3524. In this manner, the high-voltage booster is controlled by the op amp feedback action. **Warning:** This circuit outputs a deadly potential many times higher than the lethal minimum. Extreme caution is mandated in circuit design, testing, and usage.—"Linear Applications Databook." Courtesy of National Semiconductor Corporation, Santa Clara, California.

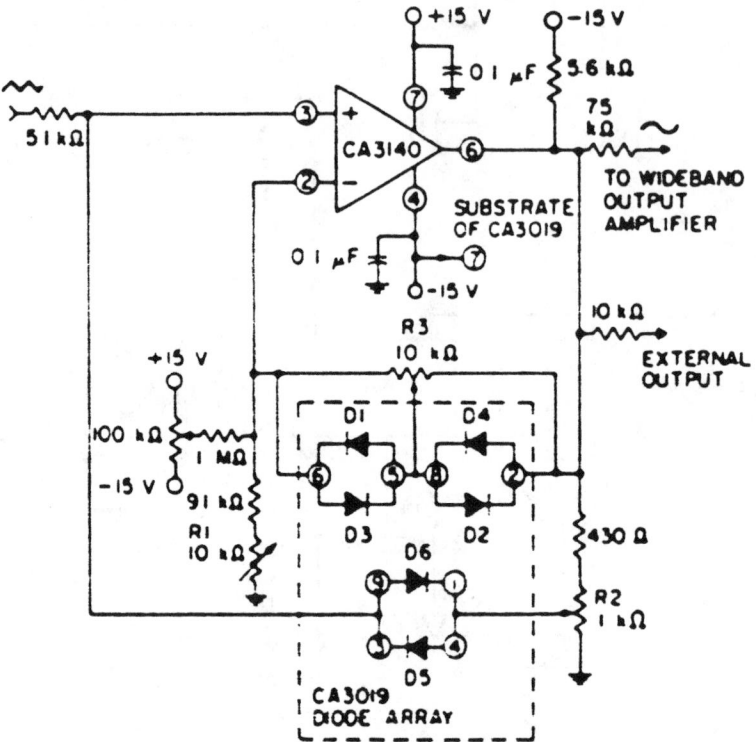

Sine Wave Shaper.   This circuit uses an op amp in voltage-follower
configuration and combines it with diodes from the RCA CA3019 array to
convert the triangular output of the function generator to a sine wave output
having good distortion characteristics. Adjustment requires a distortion
analyzer and involves setting the initial slope with R1, followed by an
adjustment to R2. The final slope is established by adjusting R3. The controls
will interact, so some readjustment will be necessary.—"RCA Solid State
Databook—Integrated Circuits for Linear Applications." Courtesy of Harris
Semiconductor.

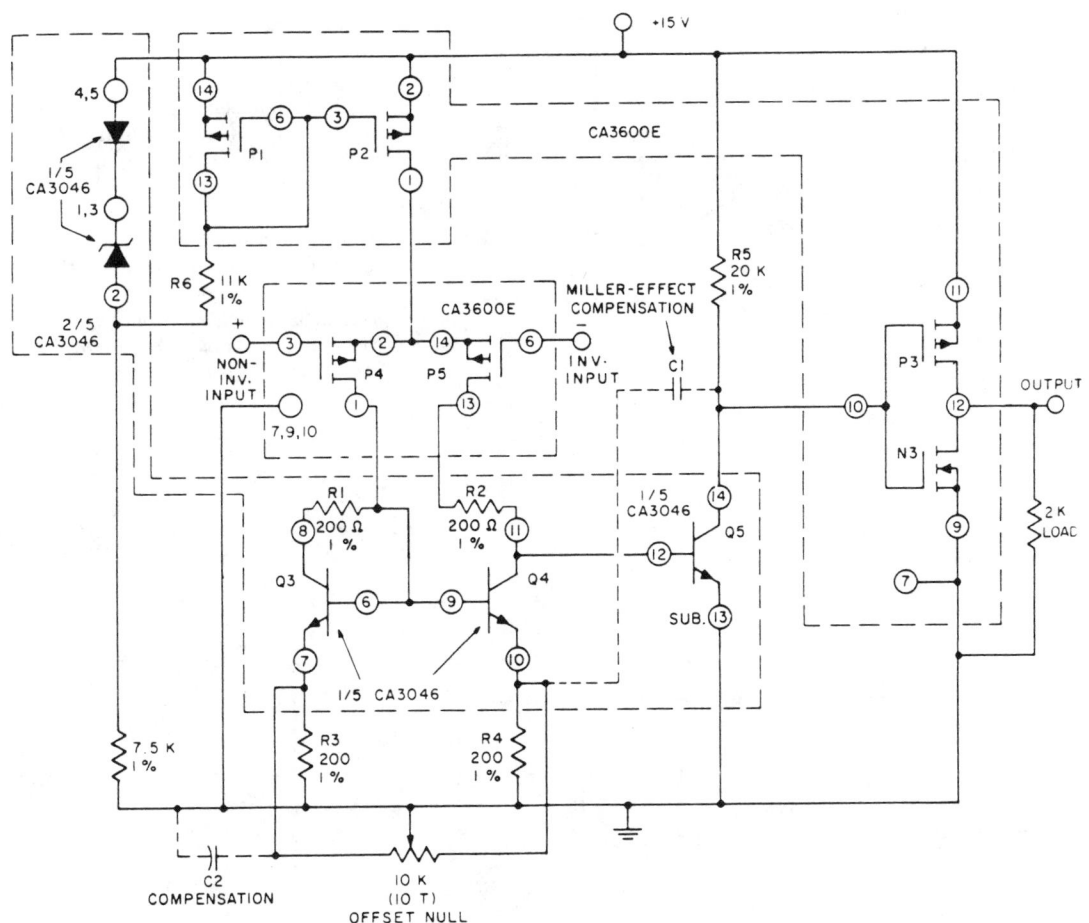

Constructing an Op Amp from CMOS Transistor-Pairs. CMOS transistor-pairs can be used in conjunction with a bipolar transistor-array IC to construct an op amp as is demonstrated by this circuit. It is particularly suited for single-supply operation and the ouput stage may be driven within 1 mV of ground potential. The open-loop slew rate is about 30 V/μsec. This circuit is particularly suited to use as a unity gain follower. It has three stages consisting of a differential input using two p-channel transistors; an N-P-N middle stage and a CMOS transistor pair serving as the output stage.—"RCA Solid State Databook—Integrated Circuits for Linear Applications." Courtesy of Harris Semiconductor.

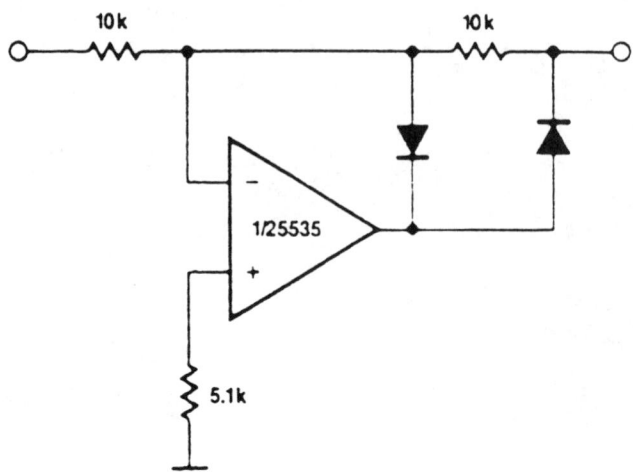

Half-Wave Rectifier. This circuit provides accurate half-wave rectification of the incoming signal and exhibits a gain of -1 for negative going signals. For positive signals, the gain is unity or 0. Polarity reversals are easily accomplished by simply reversing the polarity of the two diodes. This circuit is effective to about 10 kHz with reasonable distortion characteristics of 5% or less.—"Linear Data Manual Volume 2: Industrial." Courtesy of Signetics Company, a division of North American Philips Corporation.

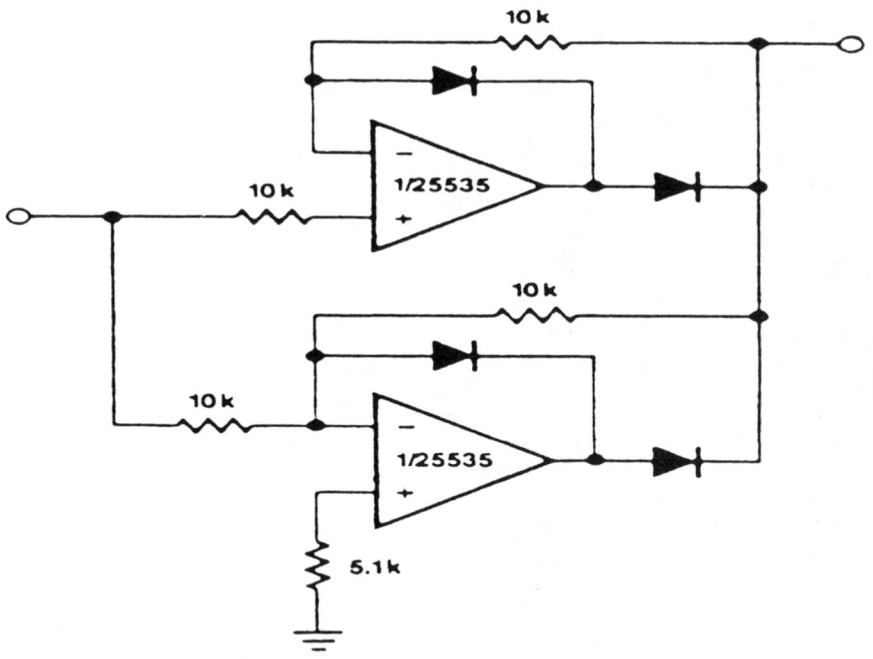

Full-Wave Rectifier. Precision full-wave rectification of the incoming signal is provided by this circuit which offers low-output resistance for both positive and negative polarities. The applied load must be referenced to a negative voltage or to ground as this circuit will not sink heavy currents. A small amount of sinking occurs across the 10-kΩ resistors.—"Linear Data Manual Volume 2: Industrial." Courtesy of Signetics Company, a division of North American Philips Corporation.

Absolute Value Full-Wave Rectifier. During positive excursions, the input signal is fed through the feedback network directly to the output. At the same time, the positive excursions of the input signal also drive the output terminal of the inverting amplifier in a negative-going excursion such that the diode effectively disconnects the amplifier from the signal path. During a negative-going input signal curve, the op amp functions as a standard inverting amplifier with a gain equal to R2/R1.—"RCA Solid State Databook—Integrated Circuits for Linear Applications." Courtesy of Harris Semiconductor.

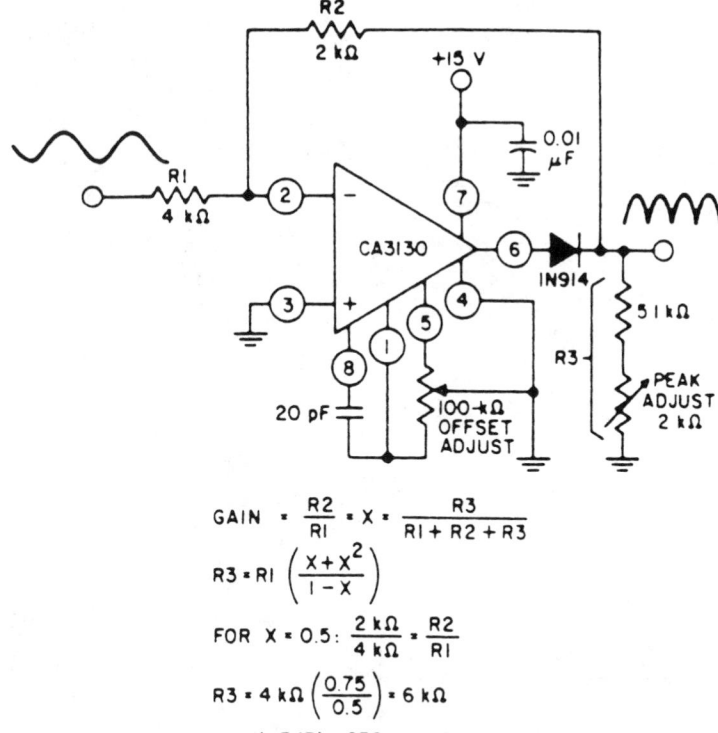

$$\text{GAIN} = \frac{R2}{R1} = X = \frac{R3}{R1 + R2 + R3}$$

$$R3 = R1 \left(\frac{X + X^2}{1 - X} \right)$$

$$\text{FOR } X = 0.5: \frac{2\,k\Omega}{4\,k\Omega} = \frac{R2}{R1}$$

$$R3 = 4\,k\Omega \left(\frac{0.75}{0.5} \right) = 6\,k\Omega$$

20 V p-p INPUT: BW(-3dB) = 230 kHz, DC OUTPUT (AVG.) = 3.2 V
1 VOLT p-p INPUT: BW(-3dB) = 130 kHz, DC OUTPUT (AVG.) = 160 mV

Capacitance Multiplier. This circuit simulates very large values of capacitance in a compact package. When C1 is roughly equivalent to a value of 10 μF, the simulated capacitance effect is 10,000 μF. R1 must be as large as is practical, because circuit Q is limited by the effective series resistance.—"Linear Data Manual Volume 2: Industrial." Courtesy of Signetics Company, a division of North American Philips Corporation.

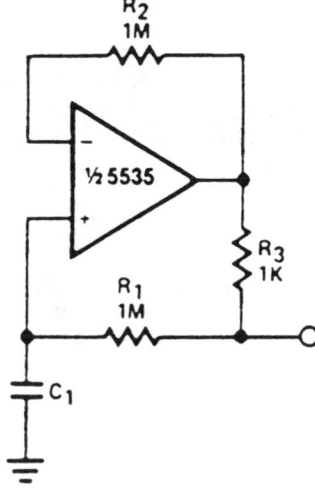

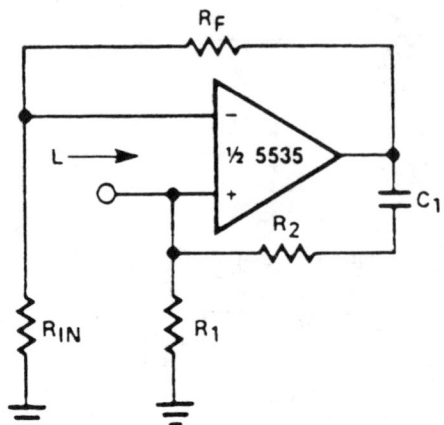

Virtual Inductor. This circuit emulates a true inductor. In other words, the output from this circuit increases with frequency. The effective inductance of this circuit is determined by the formula L = R1R2C. The Q of this virtual inductor depends upon the value of R1 being equal to R2. However, the positive and negative feedback paths are equal. This leads to the definite possibility of instability at high frequencies. To guard against this condition (more accurately, to make its presence less noticeable) make R1 slightly smaller than R2. This slight sacrifice in circuit Q is a good trade-off for increased stability.—"Linear Data Manual Volume 2: Industrial." Courtesy of Signetics Company, a division of North American Philips Corporation.

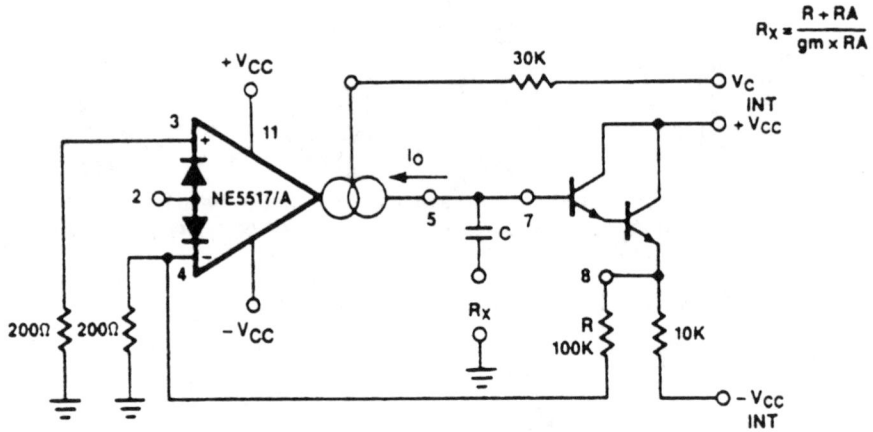

Voltage-Controlled Resistor. The voltage-controlled resistor has the capability of varying current proportional to a controlled voltage. This circuit takes advantage of the possibility to control a resistor via the transconductance of this operational transconductance amplifier.—"Linear Data Manual Volume 2: Industrial." Courtesy of Signetics Company, a division of North American Philips Corporation.

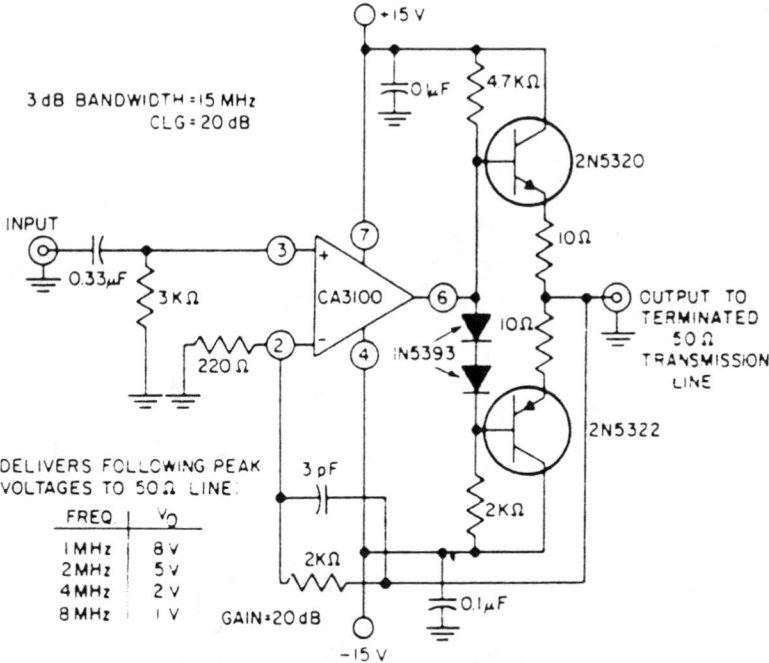

3 dB BANDWIDTH = 15 MHz
CLG = 20 dB

DELIVERS FOLLOWING PEAK
VOLTAGES TO 50 Ω LINE:

FREQ	V_o
1 MHz	8 V
2 MHz	5 V
4 MHz	2 V
8 MHz	1 V

GAIN = 20 dB

20-Decibel Video Line Driver.   The output from the wideband op amp is fed to a push/pull output stage that is designed to be terminated in a 50-Ω unbalanced transmission line. It is usable to about 8 MHz where it will deliver approximately 1 V to the 50-Ω line. At frequencies around 1 MHz, output is about 8 V.—"RCA Solid State Databook—Integrated Circuits for Linear Applications." Courtesy of Harris Semiconductor.

Circuit Index

abo- 0056

31.8